# THE GRENDEL PROJECT

## BRENT KITCHING

iUniverse, Inc.
New York   Bloomington

# The Grendel Project

This is a work of fiction. All of the characters, names, incidents, organizations, and dialogue in this novel are either the products of the author's imagination or are used fictitiously.

iUniverse books may be ordered through booksellers or by contacting:

iUniverse
1663 Liberty Drive
Bloomington, IN 47403
www.iuniverse.com
1-800-Authors (1-800-288-4677)

Because of the dynamic nature of the Internet, any Web addresses or links contained in this book may have changed since publication and may no longer be valid. The views expressed in this work are solely those of the author and do not necessarily reflect the views of the publisher, and the publisher hereby disclaims any responsibility for them.

ISBN: 978-1-4401-0449-7 (pbk)
ISBN: 978-1-4401-0450-3 (ebk)

Printed in the United States of America

iUniverse rev. date: 3/6/2009

# CHAPTER ⊕NE: THE DISC⊕VERY

Over the course of his lifetime, Professor Eric Parker had come to think of not just his students but all mankind, even perhaps all living beings, as potential points of pure light.   Each person emanated their radiance to a greater or lesser degree depending upon their level of awareness and life circumstances. In the perfect classroom, or for that matter ideal society, every moment would be bathed in the warm effulgence of pure being.   Of course he knew that his dream was not only impossible but extremely impractical.  All physical life had to confront the challenges posed by constant conflict and struggle. Nonetheless he perceived his students as possible sparks of divinity.  Most of the people whom he encountered, however, lived in the dimness of a twilight world, a perpetual grayness that suggested the onset of a spiritual as well as intellectual death.  They were potential stars, who out of ignorance and societal induced indifference, were reduced to well-appointed beggars hobbling along the shadowy streets of their damaged existences.

Parker, a professor of Comparative Religion at Durham University, had slowly adjusted to the staid atmosphere of the prestigious school.  Because each of the many buildings had a slightly different architectural design, at first glance someone new to the campus might not notice that they were all constructed with the same massive granite blocks, each intended to withstand the inevitable consequences of time. For him there was something intimidating about such grandiose uniformity, as if old ideas and theories would always be safe in the semi darkness of these damp hallways.  Unfortunately, the all encompassing grayness swallowed up any hint of individuality.

Scattered about the well-manicured grounds were statues of luminaries, great thinkers from centuries long past who were still revered although their contributions to academia were not only outdated but clearly inaccurate, disproved by decades of meticulous research.  As Parker looked down through his lancet window to the courtyard below at one such bronze tribute to the French biologist Jean Lamarck, he could not help but feel that the more informed academic communities had moved on. Yet Durham University was

1

still enamored with all of the supposed glory of classical Europe. The campus, however, managed to retain a certain charm, an impression, although clearly false, that conveyed the feeling that the university remained a bastion of integrity, even when the world around it seemed to be falling apart. To know that the school was as impervious to change as its famous granite blocks was comforting to many, particularly to a core of older, distinguished professors and administrators who were scheduled to retire within a few years. Parker, though, was not of that group; he viewed the institution's reactionary policies as unjustified, if not slightly sinister.

The handsome, middle-aged professor with a lean yet slightly muscular build was one of the most controversial figures on campus, extremely popular among liberals and free thinkers but disliked by some of the senior professors and students who were more conventional in their attitude. Nevertheless, the beginning of a new academic year was always an exciting time for him. He enjoyed returning to a campus of freshly planted flowers in the beauty of hundred-year-old gardens. He liked to think of his class and the students in it as one such garden, each mind fraught with almost magical academic possibilities. Furthermore, this semester, with the new material introduced by the Grendel Project, promised to be the most exceptional of his distinguished career.

When the class had taken their seats and quieted down, he started the projector. The eyes of each of his students were riveted, not on the large white bandage on Parker's forehead, but on the massive, dark image that hovered over them on the screen in the front of the classroom. "Some of what you are about to see, you might find unbelievable, if not shocking," he said in a calm, firm voice, his eyes darting from student to student, hoping to make contact. Parker believed that a person's eyes were a possible avenue inward, a roadway that sometimes revealed the light of their inner being, but more frequently reflected the distraction and detachment of a restless ego. Regardless, when the students' intent expressions confirmed that he had their full attention, he continued.

"Most of you are going to remember this moment for the rest of your lives, just as the majority of your parents remember vividly where they were when they watched on TV the moon landing in 1969. Some people even feel that this project could be just as revolutionary as man's first step on the lunar landscape," he said, pausing to give the class time to grasp the magnitude of the project.

"As some of you may know, I am part of a newly formed team of professionals at the university who have been selected to analyze dozens of hours of video footage. This extraordinary material was mostly shot from the field of vision of a recently discovered species of mountain primates, unquestionably our closest living relative, and perhaps what experts have called for over a century now the 'missing link.' You might recall the now widely accepted theory that mankind evolved, not directly from the apes, as many people mistakenly believed, but from a common ancestor who produced two different but related lineages. One path led to human beings, the other to supposed extinction." Once again he paused for greater dramatic effect.

"Of course the missing link was thought to be defunct for at least five million years. Certainly more research has to be performed, but what you are about to see," he said slowly, attempting to restrain his enthusiasm, "in all probability, will significantly alter man's view of who we are, how we got here and how we function. I was brought onto the team primarily to investigate that last component, particularly from the point of view of how the religious experience in mankind might have developed. In fact, the whole project puts into question just what characteristics are necessary to experience 'the sacred.' And it begs the question, does this new species, as many respected people in the field suspect, have the potential for religious encounters?"

Parker enjoyed shaking up people's belief systems with radical viewpoints, not for the perverse pleasure of creating controversy, but with the hope of actually fostering deeper thought. In the past he had to be sensitive not to go too far over the invisible line between the acceptable and objectionable, a subjective demarcation that the university tacitly endorsed. But the Grendel Project, as it was referred to by the research team, was extremely well substantiated. He felt great confidence in the credibility of the new discoveries. For this reason he presented the revolutionary findings with an energy that at times might have appeared as overzealous to the impartial observer.

"Okay, take a look." The dark figure that loomed over the room suddenly appeared to move slightly. "It's alive," exclaimed a small girl sitting closest to the professor. Immediately the class was filled with startled responses. Some of the students reacted as if they were watching a grade B science fiction movie. One male voice shouted, "Are you kidding me," while another shouted out, "Holy shit, no way, I don't believe it." A girl in the back of the room asked, trying to remain composed, "What is it Doctor Parker, a bear...or ape...or caveman?"

"It's a man in a gorilla suit, just like the fake Big Foot video that was all over TV a few years ago," another student sarcastically chimed in. "Don't be stupid. It's all a gigantic put on, made by half cracked filmmakers, like the dissecting of the Roswell alien that I saw on TV on a UFO program. Only a fool would believe this stuff."

"I certainly encourage all of you to form your own thoughts on the matter, but before you reach a conclusion, I hope you honestly evaluate the evidence.

I know it is difficult to accept a new way of possibly seeing ourselves, but please, try to be objective. We need to approach new discovers like this with an open mind. It was once believed as fact that the earth was flat, the atom was the smallest unit of matter, and that man could never walk on the moon," Parker said to the class, his voice rising with each example while his right hand dropped after each disproved theory. "Modern science, with all of its many breakthroughs, requires a new attitude towards what we once thought was impossible," he stated in a reassuring manner. Yet several boys who were clustered in the back of the room continued to loudly condemn the tape.

"Have a little patience and I will share with you what we do know," Parker hollered above the commotion. He hit pause, leaving the controversial image of the giant primate hovering over the seated students as he waited for their attention to return. Then he began to recount the events as they unfolded. "Apparently, military aircraft through the use of infrared technology spotted what they thought to be large bears or perhaps giant wild hogs high in the mountains of an unnamed eastern European country. Scientists wanted to learn more about the movement of those animals and their effect on the ecology of the region. For this project they contacted the Alaskan Game and Wildlife Department who loaned them a team of specialists who were extremely successful in tracking and tranquilizing grizzly bears. Their break-through procedure utilized a new type of camera that scanned moving objects for a specific size and shape." He quickly held up a photograph, an enlargement of the new equipment.

"This small device was a precursor to the face scans used today by government agencies in highly sensitive areas or during politically dangerous events that terrorists might target. When an object meeting the dimensions of the desired target was caught on camera, the small, easily concealed mechanism instantly triggered a dart that delivered a dose of a recently developed tranquilizer that has revolutionized the industry. Anestrx quickly and effectively can take down, even when injected into relatively small veins, mammals weighing well over a thousand pounds. Well, eventually the device was triggered and a signal noted the exact location of the take down. It was while scientists were completing this project that they stumbled on the find of a lifetime. Their discovery of a new species was a complete accident and as much of a shock to the team of zoologists as it has been to the rest of the scientific community. Although this occurred almost a decade ago, just recently a team of professionals from different academic domains was assembled to try to accurately interpret the information."

"Doctor Parker, why would anyone in their right mind believe that the missing link, as you call it, could be alive today," another student asked.

"Well, according to Doctor Aaron Price in the Anthropology Department – I am sure that some of you have read his research on ancient man – many early Stone Age cultures refer to a super large, incredibly powerful manlike species. He has provided a multitude of myths as proof. Their description of the creature has been remarkably consistent, so similar that many scientists believe that it is impossible to simply call the collection of legends coincidence. Even more important, during the last five thousand years there have been a considerable number of citings of this creature in almost every part of the planet. No doubt all of you have read about or heard about some of these encounters," he said. Several of the students were nodding their heads yes.

"Of course it is not referred to as the missing link. But just about every world culture, from Native Americans to Tibetans to regular citizens in our country, makes some mention of a parallel species. Even The Bible talks of 'a race of giants.' In fact, it refers to giants almost one hundred times and some scholars claim to have seen massive jawbones and skulls, proof in their minds of The Bible's accuracy. More often, however, the witness of the new species is generally a solitary hiker in a mountainous region who unexpectedly stumbles on the massive biped who quickly disappears, even in the most inhospitable terrain. What makes the investigation even more interesting is that the descriptions of the witnesses are remarkably similar," he said, again pausing. "Think about it! Over five thousand years of citings and there has been very little variation in the description."

"Some more recent observers have not only photographed the creature, but have made plaster moldings of its footprints. Another scientist has provided a hair that he claims belongs to the strange animal. When tested, no DNA match to any known mammal was found. Simply put, in almost every part of the world, whether in the distant past or this present century, the video taped species has been referred to or observed. In parts of Asia it is called Yeti or The Abominable Snowman while in North America it is referred to as Sasquatch, Big Foot, The Jersey Devil or The Skunk Ape. In Central America it's known as Sisimite, in China, Almas, and in Africa, The Bili Ape. Different names for the same incredible phenomena."

"Another possible explanation, again according to Price, is the animal that you see looming above you on the screen is a minor mutation of Gigantopithecus, a supposedly extinct primate that lived in China along side of early man." Parker held up a large drawing of the early primate. "The fossil evidence, although not as extensive as we would like, is fairly reliable. Even though Gigantos, as he is called, was more than twice the size of our ancestors, in some cases over ten feet tall, he was no match for mankind's more aggressive behavior. The huge, hairy creature that mostly moved on all fours was an easy target. Apparently he had a passive nature which allowed him to be hunted to the point that only a very few survived and those had to retreat to rugged areas, generally mountainous, where man's crafty brain was not enough of an advantage to guarantee success." Parker then pointed to a large globe and enumerated some of the most uninhabited areas on the planet.

"Some speculate that it passed over the Bearing Strait before that land was reclaimed by the sea some ten thousand years ago. Price theorizes that over the centuries the species evolved into bipeds."

"But how could a creature of such incredible size go undetected for all of these years?" a chubby boy in the middle of the room inquired.

"The truth of the matter is that there are tens of thousands of square miles of wilderness on our planet where if you stray from the maintained trails or roads, you will not see another human being for weeks, or in some cases, months. Parts of the American west and Canada, as well as areas of Russia and Eastern Europe, clearly fall into that category. The creature's habitat would rarely be intruded upon, and if discovered by primitive people, the ensuing interaction could easily become the source of 'myths' among indigenous tribes."

"Anyway, literature has made numerous mention of it as well. If any of you are English majors, you probably have read *Beowulf,* one of the earliest stories written in Great Britain. The hero, Beowulf, has an encounter with a gigantic primate whom the inhabitants call Grendel. Nightly the gigantic monster would feast on the warriors in the mead hall, wrecking havoc on the strongest and bravest of our kind. In reference to this two thousand year old story, researchers have taken the liberty to name the new discovery, Grendel."

"So it's a carnivore and has an appetite for flesh?" asked a girl near the back of the room.

"Let me assure you, from what we have gathered to date, Grendel does not seem to have a thirst for human blood nor has he displayed any overtly aggressive behaviors. If anything, he probably is a vegetarian."

"How can that be?"

"Some experts have suggested that the most violent of the species might have been weeded out, so to speak, and after several thousand years of continuing evolution only the most passive and secretive among them have survived. The gene pool of the few that remain might be programmed to respond with flight and withdrawal, very useful qualities when battling humans with their superior weapons. That very well could be the reason why Grendel is not extinct and why we rarely encounter him."

At this point, Parker returned to his description of the initial events. "A team of scientists from the university's Zoology Department was brought in for the purpose of classification and identification. They had no intention of either killing or incarcerating the new species. Instead, they employed the catch, tag and release method popular with many wildlife agencies. The scientists had about a three hour window to minutely record on the video that you are about to see each part of its anatomy as well as take samples of his hair, skin, nails and blood. Their next step was extremely clever; they surgically attached a state of the art micro chip camera with night vision to the creature's upper forehead." Parker held up an enlarged photograph of the new technology.

"The device was so light that he probably never noticed it. Not only could they see and hear everything that the animal encountered, but the chip also measured blood pressure, respiration rate, pulse, body temperature and, most interestingly, brain wave activity. All of these indices provided the scientists with rudimentary indications of Grendel's emotional state and quite possibly cognitive processes. This, of course, is what makes the project so revolutionary," he said as a big smile lit up his face.

"Preliminary evidence suggested that the creature was engaged in many behaviors that were very similar to humans. Some activities verged on experiences that might loosely be termed religious, or what psychologists call consciousness transformation." He paused, giving the students time to further consider the incredible implications of that statement. Then, he again clarified his primary task. "If possible, with the assistance of the team members, I am to determine whether this new species can reach states of consciousness similar to what our culture thinks of as 'the sacred.' Furthermore, I hope that the new material will help to provide at least a partial answer to that difficult and highly controversial question."

Parker pushed "play" and continued to describe the process as the video minutely recorded every part of its massive physique. Comatose yet in no way damaged, the creature's altered state allowed the scientists more than enough time to meticulously measure Grendel's entire body. He was a healthy male in late adolescence with surprisingly few scars on his body which possibly suggested avoidance of physical conflict. He was about eight and a half feet tall, about the height of the world's tallest human, and weighed over six hundred and fifty pounds. His extremely muscular body was covered with wooly, inch-long, reddish black hair. At first glance he seemed to be a cross between an ape and a human, with a retractable thumb, a foot size about twice the size of the average man, and massive shoulders extending into very long arms.

DNA research indicated, Parker carefully explained while holding up a large chart of graphs, that most of Grendel's markers were near or in the range of humans. As the camera focused on the creature's face, Parker called attention to specific characteristics. He noted its flat nose, large ears and short, sloping forehead. Most importantly, he revealed that Grendel's brain cavity was substantially larger than a gorilla's and that the internal fissures of its brain showed extensive right brain development. As he turned off the projector, he shared with the class that he was totally in awe of the physical prowess of this new species.

"Its strength and agility makes man appear to be the bearer of mostly recessive genes. What remains to be determined, however, is Grendel's mental capacity, which could have even more far reaching consequences." Attempting to restrain his excitement, he asked the class if they had any questions.

"When did all of this happen," one student asked?

"Because we are trying to protect him, some specific information has been omitted, but it was sometime around the mid nineteen nineties," Parker noted.

And then, "Is he still alive?"

"We are not sure, but the best guess is yes, he is alive."

Another student asked, "You seem to think that he is something between the animal kingdom and humanity. Why does it matter?"

"Up until now, all we could ever do was to speculate about early man. Now we have a concrete experiment from which we can directly learn. In the area of religion, to be able to understand how primitive man encountered the divine, well that would offer insight into the deepest part of our nature, perhaps revealing the very foundation of how the human spirit developed. Most of us take it for granted that ancient man was less capable than we are and that is why in conflicts with a more evolved, modern man, he died out. Simple evolution, right? It's the logic behind man's domination of the earth," he said, pointing to the blue, green globe to his left.

"But what if the missing link were more intelligent than we imagined; what if its brain evolved differently, developing some areas, perhaps, more fully than our brains? And, in the process, what if it learned to completely avoid aggressive behavior as well as contact with humans, who by comparison are incredibly violent? Many believe that if mankind could temper its acts of aggression, the planet's future would be much more secure. Of course, this is all speculation, but we hope, in time, to be able to resolve in a definitive manner all of these issues." When the class ended, many students walked out into the hallway with disbelieving looks on their faces.

A young man studying to be a veterinarian hesitated before he reached the door and then slowly turned. "You know, it's not as crazy as a lot of people might think. The East Coast mountain lion has been spotted by hundreds of people from Georgia to Maine, some even photographing it, but according to The National Wildlife Commission, it doesn't exist. If a hundred and twenty pound mammal can go unverified in our country, as densely populated as it is and with all of our technology, it is very possible that a larger animal could go 'officially' undetected in other less inhabited parts of the world."

"Thanks, that's a very interesting comparison and it isn't entirely surprising," the professor said as he closed the heavy wooden door behind them. "I'll mention it to the class."

When Doctor Parker returned to his office after lunch at the student center, his custom now for the past seven years, his length of employment at Durham University, he found a large package next to his computer. He quickly ripped open the manila envelope and hurriedly dumped onto his desk the contents. He couldn't wait to view "Grendel Experiment II" and the nearly hundred pages of written material that accompanied it. The documents included two sets of analysis, one by Doctor Price from the Anthropology Department and the other from Helena Bronsky, the university's renowned psychology professor. Each had provided a detailed account of their interpretations of the remarkable footage. With one swoop of his hand the material was safely packed into his briefcase, and seconds later he was on his way to the parking lot. Normally Parker, who rarely acted impulsively, would not leave before 4 p.m., his afternoon filled with student conferences, the grading of papers, and occasionally, technical discussions with other religion professors.

As he drove home, Parker thought about the unprecedented implications caused by recent events. He recalled the clandestine meeting that the three professors had with a team of scientists from the university's Zoology Department and the remarkable news of the discovery of a new species of primates. Although clearly intrigued by the irrefutable evidence, proof of what he and others were calling the missing link, he was most interested not about the extraordinary anatomical specifications, but the implications of these findings on life in the twenty-first century. He wondered *how did Grendel, as he now fondly thought of the creature, not only think but conduct his day to day existence? More precisely, how "human" was he? Did he possess, as some suspected, the capacity to change his level of consciousness, a quality that many scientists traditionally cited as the ultimate distinguishing characteristics that separated humanity from the rest of the animal kingdom?*

As he pulled into his driveway, he saw the unnatural smudges of redness still faintly visible on the hot concrete. His mind, still seeking resolution, immediately replayed the ugly incident once again: his wife finding him slumped next to the car door, blood gushing from his forehead. He remembered the voice in the darkness that shouted "no more," the frantic trip to the hospital, and the row of uneven stitches. Now everything rushed through his brain like hazy images from a poorly made low grade movie. Yet it was his life, his reality, as absurd as it seemed.

9

Fortunately it was not a bullet, but a small stone hurled with amazing accuracy or propelled from a sling shot, a weapon still popular in some regions of the Middle East. The police found a few suspicious footprints, size ten sandals they thought, and nothing more. The investigation was labeled "unpromising" and soon ended in the cold case files. Parker spent days searching his mind for answers. *Who had he offended? What was he to stop doing?* Like most semi-public figures, occasionally he received hate mail. There had been a few nasty letters sent to him during the past month. Recently he had erased a message on his phone answering machine that demanded he "stop degrading God." Or was it Allah? Words like "blaspheme" and "defamation" were used, but not directed to a specific reference point that had meaning to him. Once or twice Parker asked himself *was the problem the Grendel Project?*

During the past spring he had appeared on TV, part of a panel exploring "possible bridges" to the religious divide. While on that program he called all forms of religious fanaticism misguided, if not outright evil, and begged moderate Muslims, in particular, to reject the extremists and to publicly condemn those endorsing jihad. When asked by the mullah, an Islamic religious teacher with a long, black beard and a dark turban wrapped around his head, why moderate Americans had not called for a tempering of the outrageous greed promoted by the United States and its glorification of capitalism, Parker had no response. The mullah was strongly critical of the secular values that American movies and advertisements were promoting. Even worse, he harshly condemned the pornography that the US was exporting to the rest of the world. The real evil, he asserted, was America's vile degradation of everything sacred. The western media, intended or not, was primarily to blame. The mullah calmly stated that this type of satanic corruption could be ended **only** when enough righteous people opposed the filth and "shameful" exploitation of women that western cultures clearly condoned. The mullah finished his tirade with references to America's desensitization and, even worse, penchant for perversion.

Last month Parker had appeared on Fox News, and as a spokesman for the Grendel Project, introduced to the world their recently discovered new species, the proverbial "missing link" He kindly referred to it not as a subhuman creature, but simply as "Grendel." Then he shared some of the fantastic footage of the early video tapes, including Grendel's extraordinary physical dimensions. He concluded his conspicuously upbeat presentation with speculation about the possibility of a completely new interpretation regarding the development of primal man. The vast majority of the worldwide response was positive, with phrases like "revolutionary discovery" and "breakthrough interpretation" used frequently. He was pleased with his new acclaim.

A few people, however, were outraged; they offered no evidence but simply maintained that his findings were bogus. They were most offended by his suggestions that mankind might have a "peer" and that science's search for intelligent life in the universe need to extend no further than our own backyard. To them, these claims were nothing more than "shameful propaganda to undermine established religion." He had no right to challenge the teachings of God or Allah.

Parker knew that new belief systems always encountered violent opposition by traditional institutions, but he thought *this was the twenty-first century*. Besides, he had irrefutable DNA evidence and years of meticulously documented scientific material. This was not like the Loch Ness monster dispute where no physical body had been handed over to scientists for analysis. Nor was it like the UFO craze where no extraterrestrial space ship had been located and systematically examined. No, the proof of a newly discovered, highly evolved primate was uncontestable and available for all to evaluate. *The world doesn't have to like it*, he thought, recalling Copernicus, Bruno, Galileo and Darwin, each thinker presenting to mankind what is now considered universally accepted truths. Yet each of these heroes during their lifetime was condemned, imprisoned, or worse, burned at the stake.

As he walked to the front door, he stooped to pick a few large weeds from a nearby flower bed. He entered his house, a modest three bedroom cottage set back on a heavily wooded two acre lot, and then washed his hands in the kitchen sink. Through the window he could see the neighboring houses, but none were close enough to interfere with his feeling of complete privacy. It wasn't that he didn't like people, he just didn't like the thought that he had to live his life for the approval of others. Parker was a simple person by nature who loved the natural world, the stimulation of ideas and, most especially, his wife Kathy and their two sons, Matthew and Mark. If asked how his life could be improved, he probably would request more time at home with his family.

Once he had stated during an interview for the school newspaper that "contemporary life forces everyone, no matter how genuine their intentions, into a hypocritical, or even worse, contaminated existence." He vowed to resist, as best he could, his culture's call for affluence as a means of defining oneself.

One of his greatest pleasures was his small library, hundreds of volumes, mostly paperbacks that explored the finer points of world religions. He was interested not only in their myths and rituals, but more importantly, the pathways to the highest that each tradition proposed. His four publications were haphazardly interspersed with the other works, his voice just one of thousands in a continuous dialogue about the sacred and the secular. It was the potential of the dialogue that he found most alluring. He strongly believed that the scholarly process, what he had devoted his life to, included every point of view and was always open to revision if new interpretations justified new positions. Most of his fellow professors thought of him as perhaps a bit "out there," but at the same time a totally open-minded seeker, devoted not to a particular cause or religion but to the unbiased pursuit of truth.

Yet Parker sought more. On one level he was an idealist, constantly entertaining thoughts about the highest expression of mankind's true self, but he wanted something far deeper. Personally he desired to bathe in the pure light and to taste the sparkling water at the top of the proverbial spiritual mountain. On many levels, he found that book-learning was insufficient. He wanted to directly experience, not just read about, an elevated existence. He agreed with Emerson's quotation, "For most people, there is not enough life in a lifetime," and he was determined to live his life on the highest level. His driving purpose was to cultivate the powerful light of his inner being, not just go through the motions of carrying out the hand-me-down instructions of a largely misguided culture.

Nevertheless, he was always humble, never judging others by the standards of his uncommon lifestyle. For him, experiencing joyfulness in simplicity, revering the natural world, and living with an uncluttered mind, these were the qualities that constituted a high quality existence. Buddha-nature and Christ-consciousness, both never fully attainable in ordinary reality, were real goals that he challenged himself to try to emulate. His daily half hour morning meditation, a quiet period of cleansing, helped him to remain centered and at peace with himself even in the "publish or perish" madness of academic life. Some people, behind his back, made jokes about the futility of his quest. Parker would have totally agreed with them. Total success in any spiritual endeavor was impossible. But the dream of a rich internal life consumed him. He wanted each day to be an unwavering point of light, one small candle shining against the intervening darkness.

Contradiction, however, characterized his world. He was rooted in the nine-to-five work day that required him to play the game. He wanted all of his actions to be genuine and virtuous, to live with total integrity, but that was impossible in the highly charged political climate of his university. He endured because he had no choice. He was a breadwinner and partly responsible to provide for Kathy, Matt and Mark. Their needs, unlike his, were largely shaped by the distorted images fabricated by the media. Because they wanted what our culture had to sell, he had to sacrifice many of his personal convictions for the happiness of his family. It was difficult, but he tolerated Kathy's growing number of unnecessary purchases.

For Parker, compromise was an ugly word; nevertheless, his reality was formed by it. From his point of view, authentic living could only occur **outside** the boundaries of social conventions. Everything else was a performance, often perhaps subconscious, to gain material wealth and approval from others. He realized that social expectations kept everything in place, but likewise, they offered no reliable meaning regarding the true intentions of the people and events that surrounded him. Required behavior was false and could never be trusted. Yet in mid life, he could not completely escape the charades that he deeply hated. He helplessly watched his sons as they were molded by the clever aberrations of TV reality, as well as Kathy's unconscious need to be a super mom. Like most of her friends, she had developed an ever greater dependence on material consumption which she perceived as an acceptable way to define herself and her family. Parker knew that it was useless to intervene. He loved them, but gradually he spent more and more time cultivating his inner garden and its hoped for wisdom. Unknowingly, it was that serene spiritual center, the constant seeking of the light, that most occupied his being.

His early years with Kathy were largely outside the context of the shallow cultural norms that he detested. They lived in a small, rented farmhouse, with garden-picked meals in the summer and toasty evenings by the fireplace in the winter. Kathy played the flute, wrote beautiful poems and kept up with his intellectual journey. Everyday they communicated as fully as possible the glow of their inner lives. They knew that their love for each other was truly special and gladly shared the daily chores as a way of spending every free moment together. They felt liberated, happily carrying out their simple activities, seemingly beyond time and duties.

The birth of their sons and his promotion to the highly regarded Durham University, each a dream come true, forever changed the fabric of their lives. Any desire to live as a true Christian or Buddhist could only partly be fulfilled by moments of meditation before the day began or hours of religious reading as the day ended. His many duties, however, weighed on him. As often as possible, he sought that special inner garden of effulgent light. Regardless of the many roles that he had to play, he always wholeheartedly reached for something more. Unexpectedly, in ways beyond his wildest imagination, the Grendel Project brought together many of the threads of his yearning. In Grendel he encountered a living version of natural man, what he thought of as the pure energy at the core of each person's life. Now he had right at his fingertips what he speculated could be the embodiment of God's intention for mankind and a completely new paradigm for understanding divinity.

He poured a drink of water and before settling down in his home office, he went to the adjoining bathroom and removed the large bandage from his forehead. A light stain of blood seemed to continuously ooze onto the white gauze. The wound was ugly, a jagged red mass of eight or nine stitch marks that resisted healing and often throbbed. Once cleaned and bandaged again, he reached for the video marked Grendal Tapes II, Late Summer/ Early Fall, 1995. *Why did the zoologists wait so long to share the tapes with the academic community? Was it largely based on a concern for the new species' safety as they claimed?* But before he scrutinized the real evidence, he decided to read some of the reports.

Doctor Price, who had already offered several hypotheses, put forth another possible explanation, observing that Grendel might just be a larger version of Neanderthal man. He noted that most of Grendel's anatomical features were consistent with this earlier primate. Numerous experts claimed that Neanderthal was more robust than modern man and they used phrases like "bigger, thicker, stronger, much more powerful" to describe him. Some even noted that, like Grendel, he had a large brain cavity. All of these qualities were favorable for the continuation of the species. Price noted that many other "extinct" animals had managed to survive, so why not an intelligent, adaptive, close relative of ours? He concluded that since most cultures have felt the presence of a massive but elusive creature and that the accounts of the Grendel phenomena go back to the beginning of recorded history, there was certainly a moderate probability of his existence.

More importantly, Price provided evidence that even with our most sophisticated technologies, many creatures managed to thrive outside the scope of man's radar. Perhaps the most famous discovery occurred accidentally in the pursuit of the Loch Ness monster. Although no irrefutable evidence for "Nessie" existed, a prehistoric fish, known only through its fossils and thought to be extinct for fifteen million years, was recently discovered in the deep, frigid waters of the lake. Furthermore, every year, all over the planet, fishermen have found in their catch previously unrecorded species, some no bigger than a man's fingernail while others weighed over a ton. Most reasonable scientists maintained that it was indisputable that species, some amazingly large, had gone undetected **and still** go undetected. In fact, as incredible as it may seem, some scientists estimated that at most ninety percent of the species on this planet have been identified. Parker remembered that the platypus, an entirely new animal that was part bird and part mammal, was not discovered until the middle of the twentieth century.

Price's notes also included another interesting fact. Many Indian tribes on the west coast had legends regarding a Grendel type creature that they called "The Hairy Man." What most fascinated Parker were their pictographs, which included the creature. In one depiction, probably no more than a thousand years old, "The Hairy Man" was standing on the prairie next to a buffalo and a moose. He towered over both of the other animals. Price noted that the art work of these early tribes tended to be painstakingly literal and therefore remarkably accurate. They only depicted what they actually knew existed, unlike more sophisticated societies that sometimes took the liberty of exaggerating physical reality.

Parker found Doctor Bronsky's impressions of the video even more interesting. She concluded that Grendel was "highly intelligent." The chairperson for the Psychology Department and author of numerous books on the development of the brain, including her best seller, *Right Brain Gifts in a Left Brain Society*, was reasonably certain that the new species was surprisingly adaptive, using its sharp sense of smell to evade encounters with human beings and other aggressive animals. Even more astounding, Grendel seemed to **understand the purpose** behind many of man's activities. She believed that animal intelligence had been consistently devalued by man. Since we used thinking and problem solving as a measure to distance ourselves from the rest of life on this planet, the notion of a "thinking" animal threatened man's sense of superiority.

Parker's thoughts immediately shifted to Giordano Bruno who was burnt alive by the Roman Catholic Church for his assertions that the Earth was not the center of the universe and that mankind was not the only intelligent life form in the cosmos. *That was only four centuries ago,* he thought, *yet fundamentalist teachings still insisted on these outdated doctrines.* Many world religions today still contended that man was vastly superior to all other life forms on this planet, if not the universe. Because we were made in the image and likeness of God, we must possess more evolved brains and, something even more significant, a consciousness, which most theologians referred to as a soul. The majority of experts, both scientists and people of faith, contended that animals lacked possession of a divine spiritual energy, a special quality that they claimed man alone received as his birthright. Parker always felt that this assessment was unfortunate. In his mind all life was endowed with a sacred light, and therefore, on some level Godly.

Bronsky's account provided ample documentation that the brain of many mammals utilized both logic and problem solving skills. A dog recently dialed 9-1-1 with its nose when his owner was injured and disabled. Chimps had constructed both tools and weapons, devices designed to make their lives easier and more comfortable. Furthermore, some had learned a vocabulary of over a thousand symbols with which they communicated with humans and other chimps. She claimed that animals also possessed empathy, the capacity to understand what another living being was experiencing, which according to theologians is the basis of moral conduct. Parker remembered accounts of dolphins saving drowning people and dogs locating lost children, as well as dragging their unconscious masters from burning buildings.

All this information was impressive, but Bronsky was most interested in the special mental capacities, "intuitive ways of knowing" she called them, that some animals possessed. One dog had the ability to identify cancer in human beings. In another instance, a cat at a nursing home daily sat with each of the dozen residents. When he refused to leave the bedside of a particular patient, the nurses knew that it was time to call the relatives because that person was going to die soon. This special knowing occurred twenty eight times without an error. Based on the video, Bronsky concluded that besides obvious problem solving abilities, Grendel's brain, particularly the right hemisphere which centers on the intuitive as well as instinctive awareness, in some ways could be more evolved than a human's brain which tends to be time-locked and dulled by routine. Modern man, Bronsky claimed, spent so much time processing inordinate amounts of information, working almost exclusively with left brain processes, that he had progressively lost the ability to understand life in a non linear manner, the unique knowing that many animals seemed to possess. Clearly the transmissions showed Grendel making intelligent, value-oriented responses to the stimuli of his environment.

Although Parker was not comfortable fully rejecting the objective truths of the scientific world, he was certain that they were not entirely useful in the pursuit of understanding the Source. He believed that all true experiencing of the sacred was based on subjective knowledge, a firsthand and very personal encounter with the divine. Apparently Bronsky agreed. Direct experience was the only path that could lead to genuine spiritual light. The dogma of most religious institutions helped to provide moral order and some redemption from wrongdoing, but never the wisdom of genuine enlightenment. He knew that the darkness or illusion that so powerfully dominated most lives could never entirely be dispelled by the more objective, science-based knowledge that our culture so revered. Deeper analysis of man's ever increasing body of information, as educational systems often suggested, likewise was not the answer. Only the pure, spontaneous energy at the core of all people, not the illusionary workings of an all dominating mind, led directly to the Source. Divinity could never be achieved through a rational, formulaic approach, but rather through the spontaneous opening of our hearts to the great mysteries of the universe. Then with patient waiting, the sacred might enter. If enough people stripped away their social costumes and found the way to their inner beings, the world would be absolutely radiant, a brightly glowing energy of thousands of individual points of light merging into one glorious illumination.  For Parker, this was the ultimate dream and the only hope for the future of mankind.

# CHAPTER TWO: THE TAPES

As Parker pushed the tape into the slot, his excitement grew. The transmissions, however, required some getting used to. The minute camera and data chip that were ingenuously planted in Grendel's forehead worked perfectly. It sent on the left screen of the monitor a clear image of the creature's field of vision while the right screen was a series of numbers and blinking graphs that relayed his pulse, blood pressure and brain wave configuration. Parker was most interested in the peaks and troughs of Grendel's mental activity.

The tapes had been carefully edited, eliminating long periods of insignificant inactivity, like sleeping and foraging for food, behaviors that some zoologists might find valuable, but were largely useless for Parker's purposes. In some sections the transmissions appeared unnecessarily jumpy because all specific identifiers, names of villages, streets and stores, were removed. Some scientists might claim that the deletions compromised the project, hiding from the viewer important information that an unbiased observer needed for an objective evaluation. Parker understood the necessity, at all cost, of protecting Grendel.

He remembered as a child watching the movie "King Kong" and empathizing with the gigantic primate, especially when exploited by man. If captured alive, no doubt some crafty enterpriser would attempt to market Grendel as "the eighth wonder of the world," and maybe even attempt to have him perform tricks for the approval of a live audience. Parker shuttered at the thought of Grendel being displayed like some sideshow freak while being trucked from city to city.

The omissions gave an erratic and at times surreal quality to the transmissions, as if edited by a drunken filmmaker, but the research team felt that the integrity of the project was not compromised in any meaningful manner. Parker knew that the location, perhaps somewhere in Eastern Europe, although protected, could never be totally secure. Skin color, facial characteristics, clothing styles, all would point to a specific part of the world, but hopefully a somewhat generalized area that would provide no logical reference point for tracking Grendel.

The "Grendel Experiment II" video began from a high vantage point, perhaps a rocky outcrop half way up a steep incline or side of a cliff. Grendel, if visible at all, would appear to be part of the natural landscape, a moderately sized boulder perhaps. A small village of old, red-roofed houses was nestled into the rugged terrain. The long shadows suggested the onset of evening. Grendel, largely nocturnal in his habits, appeared to just be waking up because all of his indicators, pulse, blood pressure and brain waves, seemed below normal. His eyes slowly scanned the cobblestone streets and houses beneath him: a tired man smoking a cigarette, a young couple pushing a baby carriage, two cars slowly passing each other. He probably observed these activities many times before.

Behind a large, empty building, a barn perhaps, a fair distance from the street, erratic movement captured his attention. Three teenage males were standing above a fourth person who was on his hands and knees. The others were kicking and punching him, all the while saying things and laughing. The one on the ground, dressed differently from the others, was bleeding and unable to protect himself. The attackers methodically continued kicking and laughing. Grendel's attention was entirely focused on the beating. His head was motionless, watching. Parker's eyes shifted to Grendel's vital signs: a spike in his pulse suggested agitation or some other powerful emotion, while the short, even waves of his brain implied a cognitive activity, perhaps problem solving or meaning making. *What thoughts was Grendel associating with violence?* No doubt, animals can feel the suffering of other living beings. Parker remembered a squirrel that he accidentally hit with his car and then painfully watched in his mirror as it tried to drag its wounded body off the road. The image sent a wave of nausea into the pit of his stomach.

Grendel returned his attention to the main street, now deserted, slowly scanning up and down. *What was he looking for? Assistance for the wounded? Did he sympathize with the suffering boy as Parker had with the squirrel?* When his eyes returned to the earlier scene, he saw only the injured boy crawling in the direction of nearby houses. Parker hit pause while he jotted down a few notes. At these moments he felt more like a psychologist than a professor of religion. He concluded that Grendel's response, a strong, emotional reaction coupled with a mental stirring that suggested primitive reasoning or contemplation, might be interpreted as somewhat empathetic.

Parker found in Bronsky's analysis a section entitled "Right Brain, the Ultimate Reality," which immediately captured his interest. The psychology professor maintained that the right hemisphere of the brain reflected humanity's natural state, something that she referred to as "man's sense of being," his condition before civilization imposed its repressive structure by empowering a left brain world that constantly enforced laws and social standards. She claimed that over time, society gradually diminished each person's connection to this primal energy, a feeling that tended to root man in the moment. Eventually children inevitably constructed a picture of life that gave an unnatural power to words and ideas, and conversely, discouraged spontaneity, freedom and creativity, all right brain activities and natural drives in healthy people. The hours of play that children initially enjoyed were replaced by school and the accompanying days of labor, requiring highly adaptive, carefully ordered conduct.

Bronsky further asserted that the moral actions that emerged from the left brain in the form of cultural norms were significantly different than the virtuous acts emanating from man's sense of being. Living reasonably through the mind diminished the uncorrupted, intuitive energy that she claimed was mankind's most vital birthright. The left brain when over worked substantially hindered man's development and the actualization of his full humanity. Parker wondered, *can primates, animals that know pain and suffering, truly understand the anguish of its own kind and others that appear similar in nature?*

He remembered recently at a nearby zoo that a seven-year-old boy accidentally fell ten feet into the primate exhibit. While the boy was unconscious, a large female gorilla came over, picked him up, looked up at his helpless parents and cradled him in her arms as if her own. She protected him from several adolescent gorillas who exhibited aggressive behaviors long enough for human assistance to arrive. Apparently she felt sympathy for the injured child and exhibited behaviors that a human mother might display as well.

Bronsky cited numerous other examples of chimps forming strong bonds with humans, often trying to help a physically wounded or emotionally damaged trainer. After reading "Right Brain, the Ultimate Reality," Parker was more willing than ever to entertain the possibility that Grendel was an empathic creature with a highly evolved right brain. Apparently his response to the world around him was not learned but rather a series of natural behaviors that emanated from his deepest being. Bronsky's analysis supported Parker's own beliefs.

The next segment of the tape, possibly the following morning, revealed Grendel observing from a high vantage point the small play area of a school or Catholic church. Suddenly dozens of young boys streamed out of the backdoor, each wearing a dark blue uniform on top of a white shirt. In the jumble of movement, he first concentrated on a few boys playing tag. Parker reviewed the tape several times. *What was Grendel thinking? Was he noticing characteristics that normally do not appear in nature?* One boy was substantially overweight, although not quite obese. All of his fat jiggled as he forced himself to move. The other boy was pale and tight jointed; when he ran, his body appeared rigid, almost mechanical. Compared to Grendel's muscular physique and flexibility of motion, the children's odd movements were comical. Even Parker chuckled.

Then Grendel became preoccupied with the only adult on the playground, an average- sized man who had a stern look on his face and was dressed totally in black. From reading interviews and journals, Parker knew that this was Father John, a fictitious name given to this priest who appeared often in the Grendel tapes. The darkly dressed man angrily pointed at a small boy who was happily jumping off of a wooden railing. The terrified child immediately stopped. With a hand tightly gripped around his neck, he was led to the entrance of the building where he stood with his back to the playground. A minute later Father John extended his right arm and with much fanfare lifted a silver whistle to his mouth and then blew. All the children immediately ran to him, forming straight lines at places that he designated, according to their classes. Then dozens of identically dressed boys marched in perfect silence back into the dark, old building. Even Parker was astonished by the total submission of each young man. *Could Grendel feel the same discomfort in the obvious loss of freedom caused by the left brain demands of the institution?*

Parker wrote, "Grendel appears to be intrigued by the follies of civilized man. Could any other species voluntarily relinquish their own free will as wholly as mankind has to his authority figures?" He skimmed through Doctor Price's analysis, whether Grendel was a mutation of Gigantos or Neanderthal man, and stopped on the phrase," natural selection." The anthropology professor proposed that most cultures rewarded adaptive responses and punished individual choice, which usually was perceived as resistance, or even worse, rebellion. According to Price, human evolution in the twenty-first century had altered the meaning of "survival of the fittest." Strength, courage and cunning, once valuable for success, were now undesirable qualities. Conformity, the ability to regulate one's behavior according to the expectations of the most powerful in the group, was the essential tool for survival.

Price hypothesized that a parallel species not only existed, but endured **because** he was better suited than modern man for life in the rugged terrain of the mountainous regions of the earth. Self reliance provided an inner strength unknown to civilized man. Furthermore, Grendel's dark fur, besides providing protection and warmth, blended in perfectly with the granite crags and ledges that he inhabited. Unlike man, he was not dependent on complex social groups for education and resources, but preferred a solitary existence based on his own innate power. Only very infrequent chance encounters, both completely unpredictable and terrifying, provided any contact between the two species. Even as man encroached on more of the undesirable parts of the planet, regions like the Himalayas, Northern Cascades, Dolomites, and other difficult to access alpine territory, substantial amounts of undesirable land probably would always remain uninhabitable for humans. Price ended with the thought that, at present, no meaningful competition for this unusable land existed between Grendel and primitive human cultures.

The tape continued after sunset, in the twilight before complete darkness. It appeared that Grendel left the seclusion of the high elevation of his daytime hideout, probably in search of food and water. Suddenly he stopped moving and listened to a faint sound, a moaning of sorts. Curious, quietly he investigated. The video revealed a naked man on top of a woman, both frantically gyrating. After some grunts and slight screaming, total silence. Grendel was motionless, his eyes fixed on the couple, his pulse, which had been racing, now was returning to normal. He listened to the sound of their voices. Her words were pleading, as if asking for something impossible. The man was lying on his back, smoking a cigarette and drinking from a flask. He paid no attention to her. Her voice became more insistent, continuing to demand something that he refused to give her. She started to cry. He laughed, struggled to his feet, wobbled a bit and then stumbled towards a nearby tree. Grendel watched the man relieve himself, staring at the man's flaccid penis.

Parker noticed that Grendel's brain waves were erratic, possibly suggesting confusion. Perhaps he wondered about the purpose of their peculiar movements and then her unhappiness, although she did not seem harmed. Soon, they both struggled to put their clothes back on, falling several times, then arm in arm, as if nothing unpleasant happened, they made their way, tripping and giggling, back to the main road. Grendel followed them a short way but then lost interest. Parker could not construct any definitive meaning from this sequence. *Could Grendel distinguish physically harmful behavior from emotional suffering?* Perplexed, he concluded that his interpretation was a stretch. He reflected on the set of images and reviewed what he did know.

Grendel was a large primate, mainly nocturnal and most likely vegetarian, who existed comfortably in the inaccessible areas on the distant fringes of civilization. In all probability, in northern regions he hibernated for up to five months, thus avoiding the most inclement weather. Then he feasted on spring berries, roots and other edible vegetation. Perhaps the most intelligent non- human species on the planet, totally self- reliant and secretive, he was content living peacefully, and on occasion, secretly observing man's activities. Like a dog or some other type of higher mammal, he responded with some emotion toward his environment, determining patterns of attraction and avoidance.

Based on the sequences provided, Grendel was a passive creature who preferred solitude and had no natural enemies with the possible exception of man. Parker felt that the transmissions offered a limited but somewhat disparaging picture of humanity, portraying man's behavior as at times foolishly irrational and at other times unnecessarily over structured. Perhaps that was an accurate assessment he thought, unconsciously raising his fingers to his forehead and his senseless wound.

Three highly edited sequences, however, were hardly enough material to justly condemn mankind, yet Parker, although not fully aware of it, was beginning to develop a feeling that was more than sympathetic towards Grendel. In fact an unconscious bond of genuine brotherhood was beginning to grow inside of him. Recently, he found himself more than ever seeking the seclusion of his home, his own hideaway tucked in the heart of nature where he could safely contemplate the outside world without being consumed by all of its tawdriness.

The final segment of this tape was the most peculiar yet the most interesting. It was one long sequence of the moon, the type of repetitious imagery that Parker thought the editors would most certainly eliminate. *Why was it significant he wondered?* There was virtually no movement, just a remarkably clear picture of the full moon, its light illuminating the night sky and casting a dull glow over the landscape. *So what?*

When Parker checked Grendel's vital signs, he was surprised. His pulse was slow and steady, his blood pressure was at the lowest level of the acceptable range and his brain signals were a beautiful pattern of deep troughs called alpha waves that suggested a relaxed and possibly euphoric state of mind. *Was Grendel in a trancelike state, something like meditation in humans?* Parker listened to Grendel's breathing: rhythmic, slow and perfectly even. Over and over again, inhale, hold, exhale, hold, and then repeated, inhale, hold, exhale, hold. Each cycle was extended to almost a minute.

Amazingly, lodged on top of a rock somewhere, totally alone and engulfed in darkness, Grendel quite possibly was in a state of joyous well being. Perhaps like human meditators, he was directly experiencing in that very moment a connection in some completely non analytical way to a vague something, a wordless energy that was for him, in some mysterious manner, a unifying principle in what seemed otherwise to be nothing more than darkness. Parker imagined that if he could see Grendel's inner being that it would emit a brilliant light, far more powerful than many people's spirits. *But was that possible? Could other primates occasionally slide into higher states of being? And most important, could periods of transcendence modify the way primates, including humans, operated after their return to the repetitive routines of "ordinary" reality?* These questions pressed for an answer.

For a broader perspective, Parker read Professor Bronsky's essay, "Levels of Consciousness among Primitives." She noted that social consciousness, the dominant way of processing reality in civilized societies, was generally poorly developed in many indigenous cultures. They had a social order with some moral structure, but it was not nearly as complex as the adaptive routines of modern man. According to her study, natural man had a strong connection to both his basic, physical drives and, on some occasions, a trancelike condition called "cosmic consciousness," a state of being that culminated in ecstatic feelings of universal harmony. She explained that civilized man had become deeply attached to his ego, a highly subjective conceptualization of what constituted "the self." This illusion created the impression that every person was a separate entity, forever disconnected from the true reality, an existence that Bronsky characterized as perfect unification with the Source. Most primitive cultures empowered man's spiritual journey back into this primal state of pure awareness. She referenced medicine men, witch doctors and shamans, and the use of mind transforming drugs like peyote and mushrooms. Rituals, like hours of dancing, common to some indigenous tribes as well as the Sufi tradition, were designed to destroy the ego and encourage a mystical ascendance into Oneness.

Parker concluded that if Grendel were the "missing link," the early ancestor from whom man had evolved, shifts in consciousness, especially towards states of euphoria, would not be an unusual phenomenon. Bronsky ended her essay citing what Freud called "the Oceanic Condition," a knowing common to most children where they felt a momentary merging with a beautiful flow of universal energy, sort of an ocean of joyfulness. This sense of connection to a powerful, perfect benign Source often provided children with an innocent sense of well being, a state of grace that they gradually lost as they grew older. *Was Grendel's meditation, Parker wondered, an expression of the most fundamental bond that primates can experience, a temporary encounter with the sacred?* He knew that critics would object. Certainly the fundamentalist religious groups were sure to protest. A brain emanating alpha waves, however, clearly suggested possible transcendence, which many experts believed was an essential characteristic in the deepest religious states. Parker believed that Grendel's behavior was a strong indication that the sacred was not necessarily reserved for man alone. *Why should it be? If man, a primate, can alter his perceptions of reality by moving to higher states of consciousness, why not the possibility that other complex mammals could do the same? And even more intriguing, did animals have less difficulty slipping into these states of pure awareness than mankind?*

As Parker returned the video tape to its cardboard jacket and placed it back with the rest of the material, he noticed a sketch of Grendel, accompanied by an interview. With each new insight or bit of evidence, the complexity of Grendel's existence seemed to get more bizarre. He read the brief introduction and then the transcript. "Interview conducted by Professor Helena Bronsky, summer 2007, with an anonymous twenty- year- old Muslim male designated as KA. He is a very serious young man who is university educated; sensitive by nature, his art work is his passion and his major form of expression."

HB: I am a member of a team of scientists who are investigating the possible existence of a large, primate like species that many people in this area claimed to have seen or heard.

KA: We call him The Giant from the Darkness.

HB: I have been told that you are an artist and that almost every one of your drawings is of a creature that our team has named Grendel. A similar giant was written about almost two thousand years ago in Great Britain. Is your work only of the creature?

KA: Yes, like you, I am very interested in him as well.

HB: Why are you constantly sketching him?

KA: I am not sure. I can only say when I draw him, I feel relaxed, very peaceful, almost as nothing can hurt me...and it makes me happy.

HB: Do you think that this creature is real?

KA: Yes!

HB: Do you expect to see it again, since you must have seen it at least once before?

KA: Yes.

HB: But you don't feel any fear. Grendel doesn't scare you?

KA: Not at all. I have come to realize that drawing him is a kind of therapy. His image releases me from this screwed up world. Some people enjoy painting wildlife, you know, birds or flowers. I guess that is why I paint Grendel...it makes me happy...it is kind of an escape.

HB: How do you know what Grendel looks like?

KA: It's inside my brain. I just know!

HB: Do all of your Grendels look the same?

KA: Positioning and lighting are sometimes different, but the way he appears in my mind never changes.

HB: In your drawings he seems peaceful; I would almost call him gentle.

KA: Yes, I feel that he is very kind, definitely not wanting to hurt me, or anything for that matter.

HB: In the story that I mentioned earlier, Grendel killed hundreds of people and enjoyed it. But the way you speak of him is different. It sounds as if you see him as compassionate, almost human.

KA: Sometimes I think he is actually better than us.

HB: Can you tell me why?

KA: I feel that there is no evil inside of him...just like a deer or a bird. I get a strange sense of freedom when I draw him.

HB: I've noticed in most of your drawings that there is a circular, shiny object in his forehead. I am curious why you add that to your drawings?

KA: Because that is what I see.

HB: Are you familiar with Hinduism?

KA: No.

HB: Well in their religion they actually have holy people wearing shiny jewels in their forehead which they see as a sign of great inner awareness. Do your drawings try to suggest that about Grendel?

KA: I never thought about the meaning. It is just there...part of him.

HB: Do you ever see it as a sign or symbol that Grendel, in some ways, might be...I don't know...special?

KA: Grendel is special, but I can't tell you why. I feel it though.

HB: Thank you. Your information is very helpful.

I have purchased a drawing and enclosed a copy of it in each packet. Please note the pleasant expression on Grendel's face and the faint emanation of light coming from the "spiritual eye" in the middle of his forehead. These two characteristics were repeated in all of KA's depictions, at least two dozen in total. Truthfully, I feel an unexplained calmness when I look at the picture.

As Parker's mind struggled to make sense of all of the diverse accounts and evidence, his phone rang. Thoughtlessly, in a state of deep intellectual contemplation, he automatically raised the receiver to his ear. A detached voice calmly spoke, "No more Mr. Parker. Last warning. Next time much more blood!" Then a click of finality. *Was there a slight hint of an accent in those words?* Anxious, his heart beating wildly, his eyes automatically darted to the window and scanned the nearby woods. Nothing that he could see. He fought the feeling of helplessness that suddenly overwhelmed him. He struggled to control his fear. *Why was he being targeted?*

He realized again the sad truth of Copernicus, Galileo, Bruno and Darwin, that all individuals ultimately were vulnerable when opposing institutions built on centuries of blind tradition, especially the ones that claim to exist "in the name of God." He was one person, open mindedly seeking the truth, yet he was opposed by billions who were certain that they already knew it. *Why were they so afraid? Could new information really be that dangerous?* His pursuit, nevertheless, must continue or else all that he lived for would have no meaning.

Suddenly he could feel the throbbing in his forehead. The thin, white bandage slightly tainted with his blood was unable to offer adequate protection for an undeserved wound that continued to resist healing. As the pounding intensified, he wondered, *what else were they going to do to me, or even worse, to my family? Why do so many "good" people choose to remain lost and angry in the dark shadows of institutional ignorance? Why were so many well intended fundamentalists so utterly deprived of the grace of the universal light?*

# CHAPTER THREE: THE DIALOGUES

The following day, except for occasional throbbing from the reddish zigzag on his forehead, a nasty looking two inch scar, Professor Parker's class was proceeding as usual. He appeared relaxed and his affable self, his somewhat athletic physique comfortable in tan slacks and a colorful short sleeve shirt. Although an expert in many of the world's great religions, he was not content to simply be an academic instructor, a master of book learning only. On a personal level, he actively sought to be deeply immersed in the direct experience of the light and if possible, radiating it to the world around him. For Parker, teaching was a special form of illumination, a beginning point for the sacred journey up the mountain to where one first sees the breaking of dawn. Perhaps one day some of his students could taste the blissfulness of the sixth chakra, the internal bundle of brilliant energy that fostered the all seeing clarity of the third eye.

Through years of daily meditation and the discipline that it required, he had sharpened his sense of awareness and his ability to concentrate, even under duress. For much of the forty-five minute class session he had successfully blocked out the erratic pounding in his left temple and an unavoidable sense of mild anxiety, not just about the awkwardness of his unusual appearance but the sinister implications of the unprecedented events of the past several months. For the first time in his life he was troubled by a feeling of vulnerability that he could not entirely shake. Unexpectedly, a dark mood could sweep over him at any moment, jolting him from his sense of well being. Nevertheless, he conducted his class with quiet confidence, using the Socratic process to elicit responses from his students. He was fully engaged and genuinely convinced that his technique of questioning could lead his students to a deeper understanding of important life issues.

"So, just where do these reality pictures, which we all apparently have, come from?" the professor asked as if he had no clue.

"Everything we see, hear and experience has an impact on us and makes us who we are," answered a boy in the back of the room.

"And is this unique set of experiences what people call 'the self,'?" Parker asked.

"'Yes, 'the self' is formed by all of that…and probably more, but I am not sure what?" the student responded.

"So from the many events of our lives, we all form a conceptualization of who we 'think' we are. And this defining moment happens at a fairly early age, does it not?"

"Doctor Parker, you make this sound like each of us simply selects our own identity, like we can control who we become," a girl near the front of the class added in a voice that suggested total disbelief.

"Is that not true? Who we 'think we are' is our personal and highly subjective interpretation of our life story as we **choose** to understand it. It is not real in any absolute sense, is it?" Parker asked with a smile. "But in time this conceptualization eventually hardens into our life script, shaping everything that happens to us. It becomes the **only** way that we can make sense out of our reality. What I find most interesting is that studies indicate that there is very little modification of this 'idea' of self after our college years."

"That's pretty depressing," the girl added, "to think that we don't change that much over the duration of a lifetime."

Another girl stated, "I don't know whether I believe that. How did people come up with that idea?"

"Well, for instance, if we 'think' we are the hero of our story, we selectively filter our reality, I guess you could say emphasizing what we want and likewise forgetting what we want, so we understand our lives in a heroic manner. No other explanation could make sense to us, so we live as if our myth were true, almost as if it were a universally accepted belief."

"If we can choose our self image, why do so many people choose negative scripts that cause so much pain for themselves? It doesn't make any sense to me," the boy in the back of the room asserted.

"We all bang up against the real world, whatever that is, and often we don't fully succeed in our endeavors. That sense of failure, messages like 'I'm not pretty enough' or 'I'm not wealthy enough,' can accumulate into a gloomy self image. We have all seen insecure beautiful women and insecure wealthy people. Their sense of self is reduced **not** by actual reality, but by their 'thinking', which to them has a much stronger validity than the workings of the tangible world," Parker noted.

"So what can we do about our identity?" another girl asked.

"Well first we have to understand the way the game is played in our culture. The media with its advertisements, many of which are designed to make us feel inadequate, as well as its endless parade of beautiful people on

TV and in the movies...all contribute to our sense that in some way we are damaged," Parker firmly stated. "In fact there are more negative self images today than ever, not because people are more inadequate, but because we live in a more toxic environment that unfortunately fosters self deprecation."

"And what does all of this have to do with religion?" another boy asked.

"Before one can begin to understand religious reality, one must carefully examine the true nature of ordinary, day to day reality," Parker paused. "The Grendel tapes, for instance, have put a radical new spin on the notion of divinity and what constitutes the religious experience. In all honesty, it has disturbed many people who simply rely on a traditional reality picture... or perhaps we should say, historically based and rather convenient belief systems."

"Like Christianity," a freckled face girl chimed in, a smirk on her face.

"It is difficult for many people, regardless of their religion, to realize that simply thinking something does not automatically make it true. Thinking that there is a heaven or a God shaped in man's image does not necessarily make it so. There is a major distinction here. **Thinking** something can be totally different from **being** something! In other words, **thinking** about love, reading a dictionary definition perhaps, is much different than actually **being** love."

As he spoke, there was another voice in the back of his head providing a much more personal commentary. *Was his script: Seeker of Great Truths? Had he lost connection with much of the world, especially Kathy and his sons, because of his "noble calling"? As much as he loved them, was "his story" destined to be that of a mystic, or even worse, a martyr? Was he living a convenient myth, a lie perhaps, where he saw himself as the gatekeeper to the great mysteries of the universe? Was the foundation upon which he was building his life just a series of 'thoughts' and no more valid than those he was calling into question?*

And at the same moment, running counter to his own insecurities were questions about the nature of traditional religion. *Did the church through its rigid doctrines put people in dark spiritual boxes, obscuring their inner eye and blindfolding their souls? Could people preoccupied with sin and redemption intuitively experience a joyous, life affirming divinity, a harmonious energy unifying all of existence? Does the "straight and narrow road" of fundamentalism superficially advocate tolerance of difference, but inevitably generate massive discord with those traveling a different path?* Unconsciously he touched his forehead, not the potent sixth chakra with its many mystical powers, but his ugly wound, the uneven lump, stitched up by modern advancements, but still painfully sensitive.

"In any event, by the time we reach the mid-twenties, our reality picture evolves very slowly, if at all. Our conceptualizations about our self and our world gain immense power. 'I'm good with numbers,' 'I like staying up late,' "I'm artistic,' and thousands of other thoughts come together to form who we believe we are. We become attached to our identity, which we carry around with us as if it were an object. Does this sound accurate to you? Does this describe most of the people you know?" he asked.

"Well for the point of argument, let's say that it is true," the boy in the back of the room chimed in. "So what? What's wrong with having an identity? We all are going to be something, right?"

"The problem is that with the passing of years our realities often feel as if they are shutting down, getting smaller and smaller. After a while for a large number of people most days seem like nothing more than a slight variation on a perpetual rerun. Perhaps some of you are already feeling this. It is here that our identities work against us, becoming conceptual cages, if not actual coffins, that often severely limit our existences," he said with concern in his voice.

"If that is true, why would any reasonable person permit that to happen in their life?" another boy asked.

"For a diminished life our culture provides a wonderful form of compensation: consumption. The loss of excitement over our job or marriage can be partially offset by a new car, a bigger house, a more extravagant vacation or, ironically enough, a life saving belief system, often in the form of a popular philosophy, like materialism, or a conventional religion, like Christianity or Islam. We learn to look for our inner fulfillment in the external world, with both our purchases as well as our place of worship. Yet all the while we neglect our being," Parker asserted firmly.

"But isn't a religion, by definition, an inner experience. There is nothing tangible so it has to be inside of a person," the freckled face girl countered. Others in the class agreed with her.

"Most religious people see themselves as good and God-fearing," Parker noted. "But is it accurate? If a person submits to the required beliefs and structures of an institution, and they do this in the name of God or Allah or some other term for the highest, are they **really** religious people? This is where the question gets difficult. Some critics believe that religion is just another form of 'happiness consumption,' an external 'fix' that adds to our 'feel good' reality. Lenin, I believe, called it the 'opiate of the masses.' I think we need to ask, 'does conventional religion **automatically** create a sacred experience?' And if the answer is 'no,' as many would claim, have the numerous reality pictures associated with institutionalized religion lost some of their luster as well as credibility? Is that why older theological traditions don't want millions of their believers to re-evaluate the nature of the religious process?"

The freckled face girl's hand suddenly shot up and attempting to repress her irritation, she asked, "How has religion lost its credibility?"

"In a time of better educated people, many see inconsistencies, if not outright contradictions, that require a new definition of what it means to be religious. For instance, is the Source a male father figure or is there actually a real Satan, or a literal, physical heaven? Questions like these have chipped away at some of the more comforting beliefs of older faiths," Parker calmly stated.

*Had he, like those he was referring to, replaced life with a sugar coated belief system, one that offered the illusion of unlimited happiness? Or, somehow was he really different?* He lived with no externally imposed dogma, no set of rules that someone else created. His path was a journey into his soul, a quest to understand and embrace all that was there, to open as many internal sacred gates as possible. His beliefs were an extension of **his** inner experience. He wasn't shutting down his relationship with the moment, a victim trapped in a suffocating reality system. No, on occasion he had made direct contact with the Source. He had been to the top of the mountain, and he not only felt, but made contact with and basked in the pure light of unrestrained divinity. In these moments he lived with an elevated consciousness. This was not happiness consumption, something that he was acquiring outside of himself, but rather the internal cultivation of self realization, the bliss of true transcendence, an experience far removed from the normal workings of the average church, temple or mosque.

"What is even more problematic is what many theologians from a multitude of persuasions have termed the unavoidable paradox, that salvation or liberation can only occur through the steadfast commitment to a particular set of doctrines or rituals. Many religions claim that they respect other faiths, but when pressed, they usually claim that there is one, fundamentally correct way to encounter the divine, their way. So, how do you handle the question, 'one path or many paths to God?'?"

"If everybody believes their way is the right way, then why discuss it?" a boy asked.

"First, not everybody believes that. Some religions, Hinduism for example, recognize other divinities than their own. But here is an even more intriguing question: 'is experiencing the sacred inherently different from being a good Muslim, or Catholic, or Hindu, or Presbyterian? When they embrace the divine, is it essentially the same experience? What do you think? Or are they intrinsically different?"

"Of course they are different," offered the freckled face girl. "There is only one Bible, one word of God, one Christ, and one way."

"Well tell me, you're Erin, correct, could not all of the major world religions say the same thing about their faith?" Parker asked with a slight smile on his face.

"There is no definitive answer to the question," stated the boy in the back of the room.

"The real question might be are these conceptualizations toxic? Do they unknowingly lead their followers to diminished realities, spiritual dead ends if you will, that often justify inhumane lifestyles?"

"That's ridiculous! How can salvation, life ever-lasting, be a dead end?" Erin asked as if the answer was self evident.

"Clearly, there are two extremely different paradigms in evaluating the religious experience," Parker stated in a calm but firm voice. "The most common method of embracing God, the older world traditions if you will, relies on adaptive behaviors that emphasize strict conformity to agreed upon beliefs. At the other extreme is what one might call natural religion which puts the emphasis on expanding the level of one's consciousness and experiencing life through our own personal growth, through encountering our own in-born, spiritual energy. This intuitive approach to the sacred encourages constant expansion of our inner being or soul. If there is a goal for them, it might be universal love or compassion, not adherence to some dogma. Their position actually claims that the institution is a hindrance and that encountering the highest can **only** happen with individual persistence through the inner way…a type of perpetual internal discovery."

"Well maybe that is what is wrong with the world," Erin snipped.

"Conversely, many well educated people believe that our civilization, and I use that word loosely, is coming apart at the seams because of the actions of the religious mind, in particular, any form of fundamentalism." He looked directly at the young lady who proposed only Christianity as a viable avenue to God. "Sometimes we use Allah or God as a justification for the most reprehensible atrocities. Is religion, as some experts suggest, the greatest threat to peaceful coexistence on this planet at the current time? Does the whole Grendel Project and the anger with which it has been received by various institutions simply highlight the short sightedness of what might be labeled 'the religious right?' These are just some of the questions, or perhaps we should call them dilemmas, that we hope to address."

As he spoke, a surrealistic image popped into his mind. He saw Grendel frantically running from a group of old men dressed in ornate religious attire; the patriarchs of the church were throwing at him rusty old crosses and broken, golden crucifixes, as if lobbing hand grenades at King Kong, and calling him the son of Satan. *Why must a creature who was super large and very different be thought of as evil? Maybe it was best to run after all,* Parker thought.

*Perhaps it is foolish to stand up to them in the name of truth.* Yet that was just the point. Could he be comfortable in a world of total conformity, where uniqueness was outlawed and hunted down, where the sacred light, if it did manage to shine, could exist only hidden in individual souls, far removed from the supposed rewards of normality?

Doctor Parker ended his thought provoking remarks with a warning. "Historically, a course of this nature, in fact the whole Religion Department, has been looked upon as merely academic fluff, as a domain that has very little to do with the 'real' world. At one time that argument might have had some merit. Most of you, in fact, are probably taking this course solely to fulfill a requirement. But how the future unfolds…whether in fact life as we know it will even continue on this planet…quite possibly will be determined not so much by governments, political agendas, or scientific breakthroughs, as most of us have been taught to believe, but by how the issues that we are exploring during this semester play out in the years that follow. Life on Earth has always been marred by conflict, and more often than not, divisions have formed along the lines of ethnic groups and their shared religious convictions. From man's first appearance on this planet, martyrs have committed abominable acts in the name of their creator." Again he looked directly at Erin, pausing to drive his point home.

"Today, or perhaps in the very near future, those God fearing people, many might call them religious fanatics, might have in their hands not clubs or spears or guns or bombs even, but nuclear weapons that could signal the beginning of the end. This course and the Grendel Project might be at the cutting edge, the very spot where we need to shine the light if we are going to move beyond the darkness caused by some of the older traditions. I can only hope that a heightened awareness can help to **peacefully** resolve this problem. In a time of incredible technological progress and of unprecedented wealth and luxury, our destiny might be shaped, unfortunately, by the least enlightened among us and their execution of ancient, but perhaps not so holy, religious commandments." After the class ended, he headed for the faculty dining room satisfied with his presentation, yet wondering *how many students really cared about cultivating deeper levels of awareness.*

Later that afternoon as Parker relaxed in his office, he received an email from Steven Herder, a favorite colleague and a professor of Middle Eastern Studies, who quickly got to the point. "Your interpretations of the Grendel tapes, judging from the online chatter, have created a great deal of outrage among hard-line, Islamic fundamentalists both here in the US and around the world. They are offended that you had glorified a subhuman into something that on some level, at least as they see it, might be more worthy and sacred than mankind. They resent your negative comments on TV about Islam.

Clearly you are an enemy to their faith! Eric, you know that their religion opposes meditation, or consciousness transformation, basically everything that you are proposing. For them, prayer is the only acceptable way to reach The Creator and ritual is essential for a holy life. The world view that you are promulgating, one centered on a 'subhuman mutant,' has to be not just opposed but, as they see it, eradicated. Eric, these people must be taken seriously. I truly fear for your safety…as well as your family's. Do not make the mistake of minimizing their threats. There is no doubt that they are extremely angry and want you dead…and the sooner the better. I know for a fact that they have beheaded people for things less offensive."

A gigantic wave of darkness flooded his mind as if a dam burst somewhere in the back recesses of his brain. He suddenly felt uncomfortable in his office. *I wonder if Kathy and the boys are safe.* Parker was not unfamiliar with the mentality of religious zealots. In a strict, fundamentalist tradition there was no flexibility or multiple ways of understanding reality. There was no light but their light. Any competing belief system was blasphemous and destined for eradication. The whole process of thinking about the sacred, the very purpose of his profession, was sacrilegious.

Parker was aware that even Christianity had its problems with this type of thinking. Not too long ago a doctor who performed abortions was assassinated as he was drinking a glass of water in his kitchen. The sniper, who patiently waited for hours in the dark woods outside the doctor's house, was an evangelist who was intent on saving hundreds of unborn lives. After he was arrested without a struggle, he calmly explained why his acts were justified, all the while quoting from the Bible. In his mind he was simply the right hand of God, following the Lord's Prayer, "Thy will be done on earth as it is in heaven."

Parker knew that the born again movement was another group angered by the Grendel Project. They had spent decades and millions of dollars advancing the idea of creationism. Grendel, a species possibly mid way between the great apes and modern humanity, was indisputable proof of evolution. Now that some high schools had modified their course of studies to include the merits of special creation, the appearance of Grendel was an even greater embarrassment. All sorts of questions raced through Parker's head. *How secure were the multiple copies of the Grendel tapes? Could they be tracked down and systematically destroyed? Was not only his life, but anyone who possessed the tapes, in danger? Even more outrageous, would they go after Grendel and slaughter him in the name of God?*

He remembered reading an evangelist's comment in a letter to the editor column. "It is beyond ludicrous to think that Christ shed His sweet blood for a subhuman primate. It is absolutely preposterous to suggest Grendel has a multitude of human qualities and, therefore, should be afforded God's grace and salvation." Some conservative Jews were quick to point out that Grendel certainly was not to be included among the chosen people. No doubt, there was no shortage of Grendel detractors.

Unfortunately, Parker knew that once a person became "the mark," as apparently he had, it was unlikely that he or she would ever escape. He could only hope that the voice of reason in these revered traditions would somehow prevail. He knew it was a flimsy prospect. Yet, to walk out on his students, on a career that he loved, as well as to uproot his family, forcing Kathy to leave her job, taking the boys out of school, and then going into hiding, well that was just impossible. Of course he would take precautions, but life would have to go on pretty much as normal, risk or no risk. He thought, *what could be worse than a man killing innocent people to prove his love for his God, a God who created and supposedly loved all living beings?* How ironic, for decades he researched and taught about the great martyrs, people for whom he had tremendous admiration, and now, quite possibly, he was going to be tested. *Could he live with dignity in an insane world,* he wondered, *or was that, as he was beginning to suspect, impossible?*

As the weeks passed, he cautiously retained his sense of normalcy by varying his routes to work, switching cars with Kathy who drove to work in an entirely different direction and juggling his daily schedule. At the university his classes progressed as he anticipated. His task, as he had come to realize over the years, was not to indoctrinate minds but to expand each student's level of awareness. By exploring the major religious traditions, analyzing the hundreds of different points of view, he hoped to encourage each student to think more deeply and eventually on a personal level answer the ultimate question, "What, if anything, truly is sacred?" Perhaps a few might even become points of light, compelled by a powerful inner calling to journey toward their understanding of the highest. Nothing in human history, however, had prompted more violence then the multitude of responses to that seemingly harmless inquiry. Even in America, a free country, the dark, old traditions still held considerable power. *Would he, his family and perhaps even Grendel be added to the long list of the persecuted?*

Although he clearly favored the idea that each individual must construct his own personal journey toward divinity, Parker also admired loyal devotees to traditional religions who experienced internal transitions that led to peace within their being and unconditional love for all life. Mother Theresa, Gandhi, Thomas Merton, The Dalai Lama, to name a few, were brilliant points of light, and at the same time, loyal members of revered traditions. For Parker, Holy Scriptures, doctrines and rituals mattered little if the practitioners were not responding to reality through an illuminated state of being, one that was based on grace and included compassion and mercy.

On a personal level, he believed that no single tradition was equally suited for the disposition and needs of every person. Although he did not attend any formal religious institution, preferring the sanctity of his own personal journey, he had experienced moments of elevation through the teachings of several eastern traditions and his reading of parts of the Gospels as well as the book of Acts. His attitude about the plurality of paths to God was summed up in the quotation that "there are as many ways to the divine as there are routes to New York City."

Because he believed that a good question was one of the highest forms of learning, most of his classes ended in discussion, usually centered on students' concerns. On this late September morning, Erin, the slender, freckled face sophomore, continued to vent her frustration.

"I read the news article, 'Grendel: Man (sort of) or Monster" and I was appalled. How could you or anyone for that matter think that an animal, something clearly not human, could have religious experiences?" A few other students were nodding their heads in agreement.

Parker asked, "Here is a question for all of you. What does it really feel like to encounter the divine? Can the left brain, which works almost entirely through words and concepts, what some religions might call 'orthodox thinking,' really produce changes in consciousness or encounters with the divine?

"Of course it can. That's why we are taught to read the Bible," Erin said confidently.

"Is it possible, Erin, to read a detailed definition of compassion and understand it fully in one's intellect without ever **being** in a state of compassion? Could primitive, unschooled humans, or Grendel for that matter, without any holy books or even the thought of salvation, have moments of uncorrupted pureness, where he is simply surging energy in perfect harmony with the world around him? And is that not a workable definition of the sacred? So again it all is reduced to the question, 'Just what is Godly?'"

The bell rang and the class ended quietly, without resolution. There was no need to encourage conformity to one point of view, but at least Parker hoped that some of the students might think more deeply about these thorny issues.

Erin, however, waited patiently for Parker to attend to several other students. She fidgeted with her necklace, a modest gold cross, and stared out the window. When they were alone, she expressed her outrage, her face flushed with anger. "I can't believe that you honestly think some glorified monkey has the potential to experience God. It knows nothing about Christ, or salvation, or heaven, and certainly couldn't tell the difference between a saint and a sinner. It disturbs me that you are undermining the religious beliefs of almost everyone in your class." Her blue eyes burned with righteous indignation as she waited for Parker's response.

"Erin, this is a class in comparative religion, examining many of the world's different traditions, and for that matter, the whole question of just what does it mean to have an encounter with the sacred," Parker said, smiling gently, attempting to relieve the tension. "Some religions, Buddhism for one, don't even insist in the belief of a formal God; however, if we use states of being as a measure for the sacred or how a person is internally experiencing their world, millions of 'Godless' people are embracing the world with unconditional love and acceptance, while billions, carrying out the rituals of 'Godly' institutions, mostly through their left brain, do nothing more than give lip service to words that they never really personally experience. You seem comfortable with a conventional way of understanding God. That works for you. But it doesn't work for everyone. I only intend to open the door to some of the possible ways one might experience a sacred reality. In the end, we can only judge what is best for ourselves; each of us must walk through the door with the most light."

"Doctor Parker, I understand that there are other world traditions, and of course I don't agree with them. As a Christian, it is clear to me that animals don't have souls, nor do they have the possibility of going to heaven. What infuriates me is that you speak of some newly discovered primate, this Grendel thing, as if he were no different than man. That simply is ridiculous!"

"Well, maybe it has to do with the way you define God. If you see God as a Him and think that people are constructed in His physical image, then you are correct. Grendel and other highly evolved primates could not experience a Biblically-defined divinity. But if," he paused while raising his right hand, "if you were to experience the Highest as an energy, as they do in Taoism, perhaps the sacred could be understood in a totally different manner...where divinity is not a system of thought, but more like a state of consciousness marked by empathy, mercy and kindness. Some people might define God or the sacred as simply the act of living in complete harmony with the flow of the universe. If that is the case, the Highest and how one achieves it totally shifts, moving away from the external, scripture driven definition and replacing it with an internal, innate, soul driven process. That way, divinity is perceived to be within, and our birthright... and quite possibly something that man, as well as highly evolved primates, could experience... perhaps not in the same exact way, not with the same intensity, but on some level."

"That's the problem," snapped Erin. "According to the Bible, which is God's Word, all living things, most especially man, are evil by nature. That energy, as you call it, is responsible for everything that is wicked on this planet. We will never find salvation trusting ourselves. Our birthright, as you call it, is sinfulness and without Christ's loving intervention, His offer of salvation, we are all going to hell." Unconsciously she grabbed the cross dangling from her chain and twisted it part way around her neck.

"So a pure, perfect, loving divinity 'decided' to create an impure, tarnished species? Does that make sense to you? What animal, other than man, is even remotely capable of carrying out acts of evil? The world's greatest predators can hardly be called evil for they simply act on their genetic conditioning, a process that perhaps a Creator designed to maintain a balance in the environment. Can we apply the same reasoning to man? When we violate the order of this planet based on our socialization, the belief systems that we have been brought up with and indoctrinated by, can we call that a genetic imperative and claim that it is our basic nature? I think **not**!"

"So the Bible, and I guess you would include other holy books as well, teaches evil?"

"In the Middle East, Muslims and Jews kill each other. In Ireland, Catholics and Protestants kill each other, in some Muslim countries different factions of Islam kill each other… all carrying out 'Godly' instructions. One thing that I have come to believe is that true God consciousness is more about living through a universal unity, what some might call universal love, than the belief systems of a particular religion. One might even assert that religious injunctions that justify, in the name of God, death to non believers or different believers, are the real problem."

"I can't see how any of this Grendel stuff has anything to do with God consciousness, as you call it. That's what makes no sense to me," she said continuing to twist her chain around her neck.

"Erin, recently a scientist claimed that he could prove the existence of God through brain scans. Certain parts of the brain actually light up and turn a reddish hue when a person is directly experiencing the sacred. He demonstrated this by comparing the cerebral processes of long time meditators with people performing left brain, thinking oriented tasks. As meditators moved into deeper states of being, their brains actually lit up; however, concentrated thinking, even when centered on totally good thoughts, produced a totally different type of brain activity. Is it out of the question that an advanced primate, like Grendel, could have moments where he experiences pure awareness? Is not such a state, many theologians would call it transcendence, holy or Godly? In fact, what is more sacred?"

"What you are talking about, Doctor Parker, isn't religion but some kind of mystical bull shit that no sensible person would ever think about. Just because the brain 'lights up' doesn't mean the person knows anything about

God, sin or heaven…or right or wrong for that matter. Can you imagine how pathetic the world would be without religion?" she said, almost barking the question.

"That's just the point, Erin. Is this planet best served by the numerous conflicting sets of formal doctrines that are attempting to eradicate each other in the name of something holy, or, as bizarre as this might sound, would the world actually be better served if the majority of mankind experienced the reddish hue of illumination? To put it metaphorically, is it better to read about the light or to be a star? That, I think, is the question!"

"There are many more important questions, like the state of your own salvation. Perhaps you should begin to think about that rather than your reddish hue of illumination, whatever that might mean," she said with a heavy sigh and then turned and walked angrily from the room, convinced that Parker was a hopeless fool unquestionably destined to eternal damnation. Over the years he had had many of these types of conversations, not in the context of the Grendel Project, but regarding the question of just what constituted the divine mind. He reminded himself that his religious convictions upon entering college were very much like Erin's. It occurred to him that a real change in consciousness, one that at least partially attempts to remove the dark residue of tarnish from our window to reality, takes many years of genuine development, if it occurs at all.

# Chapter Four: National Television

The following evening Parker appeared on a Fox News Special that was advertised as "Bridging the Religious Divide." It was a long standing commitment, one that he had to honor. In the light of recent events, however, he felt extremely uncomfortable. The panel was composed of a priest, a rabbi, a mullah and Parker, who was asked by the show's moderator to represent a common ground in which all three religions might be able to find at least partial agreement.

The moderator started by reading a portion of an email from a viewer. "The religions that have evolved from Abraham, particularly the fundamentalists within their faith, are all marred by a history of unmerciful bloodshed. The followers of these faiths can all go back to the 'word' of God as a justification for their violent actions. How can one hope for meaningful dialogue when each creed is convinced that its actions are God ordained and that all other religions and Scriptures are erroneous?" Each of the theologians danced around the question, asserting that in their faith those people who understood their religion correctly could never engage in hatred or bloodshed. Each maintained the inherent peaceful nature of their tradition and by completely avoiding the question made it appear as though no dark side to their faith existed.

"Yet the language is indisputable," the moderator interjected, dissatisfied with their lack of candor. "Phrases like 'an eye for an eye,' or 'kill the infidel,' or 'I have come to set father against son,' can only be read as antagonistic rhetoric, clearly legitimatizing violent activity. Why should a faithful adherent discount those instructions in favor of more peaceful messages?"

"That's just the problem," Parker asserted, knowing that he should remain quiet but unable to restrain his strong convictions. "Violence and bloodshed are frequently referred to in both the Bible and the Koran. Some might even claim that these works could be read as glorifying 'justifiable' vengeance. It is an undeniable part of all three traditions. If God demands blood, which is often evident in many of the stories in these holy books, why should a godly man not act accordingly, and demand blood when confronting similar situations in his life? He could do so thinking that he was carrying out God's word. Obviously this has happened frequently in the past," he said as he looked at the three men of faith who seemed content not to respond.

After a short pause Parker continued. "The sad fact is that all three of these traditions have a long history of cold-blooded slaughter…from the Crusades, to the Spanish Inquisition, to the bombing of Palestinian refugee camps, to 9/11. Presently the world is dealing with religious extremism in ways never imagined. You might remember about ten years ago a brother killed his sister in the street of a major Middle East city. He calmly walked up to her and then shot her in the head, at point blank range I might add. He did this because she had defamed her family by falling in love with a non Muslim. After turning himself into the authorities, a few hours later the brother walked out of the police station totally free. Apparently he was carrying out an Islamic law which made cold-blooded murder acceptable. This, in itself, is reprehensible enough, but when these violent mandates are carried out in non Muslim cultures, peace and understanding become impossible. Fundamentalism in any religion is a threat to world stability, but clearly, I am sad to say, Islamic extremism is the biggest dividing force on the planet."

What was intended to be an informative interchange of different perspectives suddenly turned ugly. For a moment the mullah, trying to restrain his irritation, stroked his long black beard and glared at Parker. Then, he broke his silence with an angry retort, "The biggest dividing force, unquestionably, is American capitalism, which is the real religion of the west. **Greed is your God!** Money and consumption are your forms of worship. It doesn't matter to you that you are destroying older cultures all over the planet, or that your policies are leading to the depletion of the earth's natural resources. You and your people are indifferent to these facts," he said in a raised voice, as if each word were a dart intended to render Parker useless.

"You are raping the planet and destroying its people and yet you get upset when someone fights back. Doctor Parker, surely your God, if you believe in one, does not endorse how your Christian-Judaic culture has sold itself to the almighty dollar. The holy people in your country stand idly by, and say absolutely nothing," he paused, his hand in the air pointing to the heavens, "as your capitalistic values corrupt the spirit of tens of millions with your filth. Even worse, your naval blockades deny medicine to the poor, killing the bodies of tens of thousands, mostly innocent Muslim children. These things do not concern you or most of your materialistic countrymen."

"Does our indifference, as you call it, justify car bombings, decapitations and suicidal Jihadists? Is that the God that you teach your followers to support," Parker asked, trying to hold back his disdain. "It seems to me that this is the very crux of the problem: would a truly God-fearing, religious person, no matter what set of beliefs he might fall under, commit cold-blooded murder in 'the name of God or Allah?' Would not premeditated murder be a direct violation of a spiritually-evolved consciousness? In fact can there ever be a God endorsed motive for the mass slaughter of thousands of innocent and defenseless people as occurred in 9/11? When men who refer to themselves as 'holy' call for Jihad, is this truly Allah's wishes or the will of corrupt leaders seeking to exploit the downtrodden so they can destabilize a country for their own gain?"

"You never look at yourself, Doctor Parker, or the ideology that you seem to support. When your country does not allow medicine to the sick and dying children of Iraq, when you have great abundance but turn your back on the hundreds of millions starving on this planet, does this not constitute premeditated murder or destabilization as you call it? How can you support this thoroughly corrupt culture and still have the courage to present yourself as a person of good conscience? In my eyes you are nothing but a hypocrite and a liar."

The moderator attempted to intervene, but Parker, now somewhat calmer, could not restrain himself and with a less inflammatory voice noted, "This discussion is not about American business values, many of which I find just as despicable as you, but how we can heal the gap between many of the world's great religions. The real question, I believe, is 'Does the religious experience of Christians, Jews, and Muslims have anything in common, anything that might foster greater understanding and acceptance?'"

The rabbi commented somewhat nervously, "We all believe in a higher power that requires our devotion and provides each of us with moral direction. I think we all can agree on that."

"And the intention of each of our faiths is to help make a better world," the priest responded with a smile, hoping to ease the tension.

"If there is so much commonality," the moderator asked, attempting to regain control of his program, "why then is religious based violence more rampant now than ever? Not just the obvious conflict of one religion against another, but violence within religions. We seem to have different Muslim factions killing each other, fundamentalist Christians justifying 9/11 as God's disappointment with the immorality of New York City, and even Jewish settlers being forcibly displaced by their own faith. Is this the action of truly religious people, or is there something wrong or inappropriate in describing these activities as religion based?"

"What constitutes religion has been debated for thousands of years," Parker added. "Christians rejected Gnostics, people in their own faith who

43

personally internalized divinity. They condemned people who found divinity in nature, calling them pagans. They tried to convert the Native Americans, calling them heathens. I wonder if we could reach an agreement, if not a definition, regarding at least some of the essential ingredients of the religious experience?"

"Certainly we all agree that a religious person must believe in a higher power," the rabbi said.

"Of course we all advocate a universal morality...based on love and goodness," the priest added.

"But how does a person experience God and morality? Do they believe in them because they are supposed to...or is there more to it?" asked Parker, his eyes looking quizzically from one panelist to the next. "For instance, we might all claim to love our mothers, but what does that really mean? Someone might see this as a duty, part of any God-fearing child's responsibility. The Bible instructs people to 'honor' their parents, and I am familiar with similar injunctions in the Koran. Many would argue that the mindset of duty, an external expectation supported by some moral authority, is far different than actually experiencing love consciousness. It is far different than feeling a deep sensitivity and oneness with the person who brought you into the world. Does not a true religious experience require more than duty, require the actual feeling within your heart?"

"Does it really matter why a person does something good, as long as it is good?" the moderator wondered.

"Well, apparently it matters if you are a Christian," Parker added. "It is not enough for a man simply **not** to commit adultery, to just function in a morally correct fashion. The Gospels require much more: that his heart is actually clean and pure; he must truly be in a state where he could never look upon a woman as a sex object. Apparently the unconsummated desire is just as evil as the immoral action. And I think that is the very distinction that many people are questioning. Now tell me, can a truly God-infused person blow apart themselves and dozens of other people who they don't even know in the name of all that is holy and good? What kind of God would require that?"

"It is called martyrdom, Doctor Parker, and your traditions are just as seeped in it as ours," said the mullah in a steady voice. "Is there a more defining, sacred act than giving up one's life for one's Creator?"

"I think it is the perception of most of the people in the west that this sacred sacrifice to Allah, as you call it, is anything but sacred. When one exchanges their life for sixty or eighty virgins in paradise, this hardly appears as a selfless act of divinity. It might be the act of followers carrying out what has been described to them as their duty to their religion, but as we noted before, the motives behind duty are, at the very least complex, and not necessarily good or pure."

The Rabbi added, "Judaism has a very high regard for duty; it is an essential part of our religion, the centerpiece of our tradition."

"And the Nazis carrying out Hitler's decrees were doing their duty," Parker stated emotionlessly. "But is it enough to simply do what you are told or does a truly religious person actually feel in their heart the rightness of their actions?"

"It is right because it is in the Koran and Allah tells them that it is so. A person needs to know nothing more than that," asserted the mullah.

"Would it be fair to claim the distinction here is whether religion can be defined as simply a duty required by the scriptures, or must it include something more, perhaps a personal spiritual transformation," the moderator asked?

"How can we ever know what is in a person's heart," asked the priest, a somber expression on his face. "I have listened to thousands of confessions, and most often I am uncertain about the sincerity of their words. But who are any of us to judge what is in the heart of another? Each of our faiths hopes to be an avenue towards some higher calling. All we can do is map out the way and hope for the best."

"But a map that includes bombings and decapitations might be called military strategies, but from my perspective, hardly religious," Parker concluded. The moderator signaled that time was up and attempted to end this unsettling exchange of ideas on an upbeat note. "I'd like to thank our panelists for their candor. Healing the religious divide can occur only when there is an honest sharing of differences. Hopefully we have taken one more step towards understanding and in some small way reconciling religious differences. Have a good night," the moderator said in a reassuring voice and a smile on his face.

When the debate ended, there was a minute of small talk among the panelists, and then the mullah walked with Parker towards the parking lot. "You ask too much of the common believer," he said. "Let's not fool ourselves. In most cases a religion will not change a person's heart, but it will give them a direction in life, something better to live for. All we can hope for is that they conduct themselves well. That has to be enough. Most people are like sheep. They will be content in the fold with food and leadership."

Suddenly his voice changed from the gentle father imparting a lesson to his flock to a stern taskmaster issuing a grave warning. "Doctor Parker, you are inciting a very volatile situation, upsetting many very powerful and holy people with your favorable reports on this creature, this Grendel thing. The world doesn't need a new way to think about God and religion. It needs more devotion to what has already proven itself to be the one and perfect path to our Creator. It could be very unfortunate for you if you persist with this scientific nonsense that you are spewing forth."

As they separated, Parker's mind was searching for answers. *Was his warning somehow connected to the earlier attack in his driveway or was the Mullah simply referring to the possibility that the Grendel experiment, if it did not go well, might damage Parker's career in the eyes of academia?* There was no way of knowing with certainty. One thing was clear, Islam was not accepting of diversity or opposition. When Parker was about to unlock his car door, the mullah's last words were, "Please, for your sake, and the sake of your family, do not disregard me!" *Is this a threat,* Parker wondered.

As he drove home, Parker recalled several disturbing global events. Recently in Africa the Muslim press was highly critical of a local beauty pageant that had just concluded. The newspapers harshly criticized women for parading around in skimpy bathing suits, almost naked, for the pleasure of men. When a non-Muslim newspaper commented that Mohammed would no doubt have had the winner for his bride, all hell broke loose. Numerous buildings were torched, including the newspaper's offices and printing press, and over two hundred and fifty people were killed. When other countries were critical of this "inappropriate and excessive response" on the part of Islamic extremists, their concerns were met with disdain.

In another instance, a cartoonist in the Netherlands created a drawing that poked fun at Islam. For this, he was killed in cold blood as he bicycled to his office. The assailant, far from being contrite, simply noted that he was carrying out the work of Allah. The Muslim press could denigrate Jews, Christians, capitalists and Americans, but when Islam was satirized by newspaper writers in other parts of the world, death was the result. Apparently Islamists were unable to grasp the hypocritical nature of the situation.

As he drove up his driveway, he wondered, *was there a middle ground between freedom of thought and the demands of the Islamic state?* Unfortunately, he could think of none. Most certainly, tolerance and the acceptance of difference were far beyond their vision of an appropriate society. Parker recalled how some Islamic countries put homosexuals to death and punished women who wore western style clothing. Whippings were not uncommon for that matter. No, peaceful coexistence was something that could be talked about, perhaps even suggested as their goal, but in reality, he could think of no religiously diverse culture that was living in harmony with Islam.

When he entered his house, Kathy was standing by the door with an apprehensive look on her face. She had been crying. Her eyes burned into his as she attempted to control her anger. "Eric, it's happened again. While you were on TV, I got a bunch of phone calls, but there were no words. I really can't take much more of this. I know it has to be about your Grendel research. Do you really have to continue with it?"

"It's probably the most important thing that I have done in my life. I know that there is no way that I can fully prove any of my interpretations... but it's essential that we consider a different view of our ancestors," he said, as his face tightened and his eyes drifted towards the ceiling, unable to meet hers.

"It's just that I feel so vulnerable out here in the woods, with nobody near, and having to look after the boys. Honestly, sometimes I really am afraid."

"I'm sorry, Kathy," he said, giving her the most reassuring hug he could offer. "I never thought things could be like this. The last thing I want is for you and the boys to suffer, but I really don't know what choice I have."

"Do you have to offend people with your work? You could report on Grendel without pushing such a positive view of his behavior or making him out to be some near human with remarkable abilities. You really wouldn't be compromising yourself, and we would all feel much more comfortable, I'm sure," she said with a slight smile, hoping that he would agree.

"But what if Grendel is revolutionary? What if he represents a path that mankind could have taken, but didn't... maybe in some ways a better path?"

"It doesn't matter," she screamed. "We're here. We're alive. We matter!"

"Kathy, the more scientists assemble about man's history before the advent of the written word, the more brutal our species appears. I read an account the other day that suggests that tribes used to hunt people, store them alive in cells, like we might keep meat in a freezer, and then eat them, not all at once but one portion at a time, an arm one day, perhaps a leg another, as their hunger dictated. Can you imagine that? It is impossible for me to understand that level of brutality. I am certain that Grendel is not capable of such a barbaric action."

"Eric, you have to give this stuff up. You're obsessed with it. You don't get it. It doesn't really matter that much one way or the other," Kathy said looking up at his distracted eyes. "Our family is what is most important. We'll never know for sure any of this. It's all speculation...and as far as I'm concerned, it really is unnecessary."

"But what if mankind is not the high point of evolutionary development, as history books have claimed and our holy scriptures assert? That certainly would put a different spin on what things might mean. Kathy, we have been taught that God, a benevolent, all powerful father figure is in control, directing everything in the universe and that mankind is the centerpiece of His creation. Of course we want to believe this, for it puts us in a very advantageous position."

"I don't care!"

"Please hear me out Kathy." In his excitement he continued to lecture her as if she were a student. "Our technology is so advanced, our world so far elevated over other animals, it is impossible not to think of ourselves as God's chosen species. But what if good doesn't always prevail? What if our ego, and the socialization and education that naturally grow out of it, things that we have taken for granted as good, really has moved us in another direction, covering up the real nature of our innate humanity?" He paused, unaware that he was lost somewhere in his professor persona.

When he continued, he still wasn't communicating with his wife. "I know that this is mostly speculation, but there are indications that this might be true. It sounds strange, but Grendel might be more human than us. We have been brought up to be selfish and aggressive. That's all that most of us have ever known. You know, get ahead almost at any cost. We follow the model that our culture puts forth, thinking, of course, that this is what it means to be normal, but what if normal has evolved into something that is … sub human? Something that has become progressively more and more desensitized and prone to violence! That is a terrifying thought! Kathy, I can't give this up. I wish I could, but it could lead to a breakthrough, like Copernicus or Darwin, a completely new way of understanding man and his place in the universe, and just maybe, a legitimate justification for why we should change."

"Can't you see Eric, most of mankind is happy with their belief system, happy with Jesus or Buddha or Allah. It doesn't matter one bit what you discover, nothing is going to change, except our family could be destroyed in the process. If someone wants to, the four of us could be dead in a heartbeat. Don't you see that?" Her voice cracked as her left forefinger scraped against her throat, suggesting a beheading. "There is no way that we really can protect ourselves if fanatics hate you …and want you dead. I know it's hard, but you have to give up this Grendel project, if not for me…then for the boys," she begged, her eyes on the verge of tears. Any loving father had to agree, she thought.

When he embraced her again, he felt more confused than ever. He loved his wife and children yet he also longed for truth, a clear understanding of the way things are. Would he have to choose on one hand between his own flesh and blood, and on the other, the very purpose of his existence? Could he find a way to protect his family and at the same time pursue his passion? He remembered something about Einstein's life being in jeopardy, an agent from a foreign government sent to kill him, but you could hardly compare the Grendel Project to the atom bomb. *Belief systems are sacred,* he reminded himself, *but so is the truth. And Kathy had a valid point; maybe Grendel was nothing special, a little more sophisticated primate than a gorilla and nothing more. But that was the question for which he had to find the answer.* His mind raced back and forth, hoping for some sort of resolution that never surfaced.

One thing was certain: nothing in his life, no amount of education or religious training, could ever adequately prepared Parker for his present dilemma. He felt totally severed for his mind, with its years of left brain thinking, was pitted against his heart. Logic, nevertheless, still prevailed. Parker had a plan. *Doing something is better than nothing,* he thought. First, he would share the latest phone harassment with the local police and ask that they send extra patrols to his neighborhood. It might not amount to much, but it was a start. Then he would call his agent in New York City and request a delay in the distribution to book stores of his latest book, which included some Grendel material. Although he intended to continue his research, in public he would attempt to change his persona by dampening his enthusiasm for a more evolved Grendel and minimizing his voice in the debate over the nature of religion. Hopefully, distancing himself from the project would be enough. A thought occurred to him, *all change involved risk, especially when a person's view of reality was at stake.* For years he encouraged his students to have the strength to confront the unknown in their lives, to open themselves up to new ways of understanding reality; now it was his turn to struggle with uncertainty and to actually attempt to live what he had been advocating to others all of these years. He had no choice. He had to continue his journey towards the all encompassing light, even if it revealed unpleasant shadows on an unwanted landscape.

# Chapter Five: More Videotape

"Oh my God," Parker exclaimed, bolting up to a sitting position in his bed. He was totally disoriented, his night shirt warm with his sweat. Large beads of perspiration, like giant teardrops, rolled down his cheeks. His dream felt totally real; in some ways more concrete than his life. As he wiped the beads of sweat from his forehead, he was careful not to press too heavily on the remaining scar tissue. "What is it," Kathy whispered, a terrified look in her eyes. A few moments later, "just a nightmare," he said thankfully, his voice sounding as if it were coming from a distant galaxy. He took a few deep breaths to compose himself. Then feeling normal once again, he asked, "Why is innocence always slaughtered, often in the name of something good or holy?" The unanswered question seemed more pertinent than ever.

Slowly, he recounted to Kathy what he could remember of the nightmare: "Grendel is meditating peacefully in the morning sunshine, appearing far in the distance as nothing more than a motionless protrusion on a large rock outcropping. A beautiful golden glow, like you see over the heads of holy people, floats above him. A hunter or tracker appears, dressed in some kind of a military uniform. He sees Grendel and thinks he is a bear or maybe he's been stalking him all along. I'm not certain. Then, he slowly takes aim. I screamed with all my might, 'Nooo.' A terrible shot, like the universe was exploding, followed; it still reverberates in my head." Parker paused and put his hands over his ears, his eyes still lost in some dark cosmos. "Grendel falls backward, hundreds of yards into the lake below, lost forever." As Parker felt the slap of the dark water on Grendel's body, a feeling of utter hopelessness penetrated him, pressing heavily into his body, drowning his struggling being. The senselessness, no, utter stupidity of man's inability to live harmoniously with the world around him, made his head throb, a heavy, repeated pounding that felt as if he were about to come undone. Unconsciously he fingered the scar, a lumpy mass of irritated skin that still was puffy and inflamed.

Sleep would not return. In the darkness, his bedroom seemed strangely surreal, unaccustomed shadows leading nowhere, as if he were spending the night at a dilapidated motel or an old farmhouse in a foreign country. The mullah's warning reverberated in his mind: "It could be very unfortunate… very unfortunate…very unfortunate," until fear seized not only his thoughts but his very being. He saw men with rifles and swords, their faces hidden behind black cloth. He was being escorted to a large, makeshift stage. There was a camera. He realized that he was expected to admit his guilt and apologize to the world for his transgressions. Someone raised a silver blade. Suddenly his head was sliced off and tumbling on the ground. Blood was everywhere. Terrified, he could not bear to look any more. He tried to remain calm. In the blackness, his once familiar life now seemed strangely abnormal, like an out-of-place plot from a comic book that he might have read as a youngster. Thoughts of a horrible death stabbed at his brain.

In time, the gruesome images turned into dark thoughts. He realized that murder was man's most popular solution to serious conflict, the ultimate resolution. The complete elimination of one's enemy, the source of one's pain and suffering, was usually most preferable. Certainly it was much easier than attempting to live in harmony which required continuous compromise and the working through of differences. History had revealed man's penchant for actions based on the darkness of his lower consciousness. Murder, from homicide to genocide, was sanctioned by the Holy Books as far back as language extended. Conflict was everywhere!

As he examined his life, he thought, *every day each of us murders a small part of ourselves. Everywhere people, in the name of a noble cause, like making money or family survival or worshipping God for that matter, are destroying the very fabric of their humanity by numbly submitting to years of brainless indoctrination. And I am no different!* He opened his eyes, hoping to see the grayness of early dawn, but the blackness pressed heavily upon his chest, making each breath an unnatural labor. Like a dying fish gulping for air, he fought against suffocation, straining for survival.

His mind would not shut off. He was no longer addressing a class but painfully scrutinizing his own life. *Coldness permeates most of my personal interactions which allows me to treat myself and others as if they were dead objects; furthermore, I model my world around the lifeless technology that consumes me. Even worse, I allow the numbness to grow, feeding it boredom and indifference. There is no room anymore for meaningful personal creation.* He thought of Kathy who was sleeping peacefully next to him and the boys in their bunk beds down the hallway.

He was helpless and his family vulnerable. At that moment there was no dawning of a new day, no personal, life transforming epiphanies, no glimmer of hope, just depressing, unsettling realizations: *Every government, every organized religion, every system of education, limits man's growth. For them, the individual can not be trusted; man in his natural state, they no doubt believe, is flawed and gravitates toward a lazy, lower self; therefore, they think that it is the institution and only the institution that can save him.* In the dead of night, Parker's many fears, like dozens of bizarre voices with strange accents, a cacophony of indistinct but troubling sounds bombarded his head. They were impossible to control! He wanted to escape, but where? From time to time his body shuddered, long, frightening convulsions that shook his fragile psyche.

He forced himself out of bed and he started to meditate. After minutes of deep, slow breathing, his anxieties began to fall away. Slowly, he continued to inhale the light and exhale his negativity until he felt somewhat centered again. Parker's thoughts turned to his latest book, certain never to be a best seller, but whose distribution could only make matters worse. *The Lustres: The Universal Wisdom of the Mystical Traditions* had received strongly mixed reviews. The Catholic League banned it, calling the book, "a strike against the most holy," while evangelists claimed that it was nothing more than "dangerous misconceptions" and added that it was a work of which Beelzebub would be proud. Conversely, Unitarians praised it as an "inspirational guide to finding the power of God within ourselves." The New Age movement lauded it, with statements like, "Parker finds a way to convincingly reveal the great mysteries of life" and "completely groundbreaking...revolutionary new thinking." Several religious groups even requested that he speak at their weekend retreats.

Normally he would have been comfortable with the publication. However the most recent series of unsettling personal events convinced him that the glorification of mystical wisdom would further agitate fundamentalists on both sides of the ocean. Now his ideas could only add more fuel to the existing fire and more reasons for him to worry about his future.

Before the Grendel project became his obsession, Parker had identified in *The Lustres* ten conditions of the liberated spirit, traits that the mystical branches of most of the world religions seemed to agree upon and emphasize. He was deeply convinced that the vast majority of problems that existed on our planet were caused by destructive belief systems, ones promoting the darkness of selfishness, a myopic concern over one's personal salvation, and a desire for earthly power. Many religions, perhaps not by design, actually supported policies that endorsed a disproportionate amount of worldly goods ending in their backyard. He thought of all the gold in Vatican City. *How could fair-minded people be comfortable with twenty-five percent of the world's population consuming seventy-five percent of the world's materials?* When people incorporated the mystical states of being into their everyday existences, Parker claimed a major shift in perception inevitably occurred, dramatically and most often permanently changing a person's sense of reality.

*The Lustres*, Parker hoped, would provide in a clear, concise manner a pathway to a new life, one primarily experienced on a transcendent level. That new life, what some people cryptically called "The Secret," was openly embraced by a very small minority of the world's population. Yet Parker was convinced that the teachings of the mystical traditions were infused with a right brain wisdom which when fully understood was the **only hope** for a planet on the verge of destroying itself. His book artfully wove together the enigmatic voices, all profoundly intuitive, of many of the world's great traditions. The work embraced portions of Taoism, Buddhism, Hinduism, Christian Gnosticism, Sufism, the Jewish Kabbala, and the enlightenment of Aborigines and Native Americans. Through the ten states of being that the book identified, all simple qualities inherent within human nature, Parker thought that mankind, both on a personal as well as collective level, could achieve the balance and harmony necessary for a peaceful and prosperous earth. Wisdom, not knowledge, was the answer, the ultimate light of profound, inner illumination.

One of the most ironic aspects of "The Secret" was that it was totally out in the open, available to everyone, but because the desired characteristics appeared in most cultures to possess no value, no means of real power, only a very few highly aware individuals could sense its earth-shattering potential. Even more problematic, the act of realization was far less significant than the act of implementing the instructions into one's day-to-day living. If Parker followed the wisdom of the mystics, he would seriously jeopardize his career, his relationship with Kathy, his social standing and the future success of Matthew and Mark. Philosophically, he could advocate the power of the transcendent, but very infrequently could he live with integrity in this higher realm, actually choosing to act on his idealism.

Enlightenment in his culture carried a terrible price tag: isolation and insignificance. Parker knew that he was a compromised person, like the vast majority of civilized people, but he yearned for a simpler, more genuine existence. For the moment, the best he could do was to live with one foot in each world, daily compromising his passionate quest for the highest so he could reasonably provide for his family. Yet it was a painful predicament. One should not be forced to choose between the love and well-being of one's wife and children and the quest for "the truth." He thought of Christ's rejection of his roots. Then, he wondered if natural man, the one percent of the world still living in primeval societies, encountered such dilemmas. *No, he realized, their holy men are held in the highest esteem and are placed at the very top of their culture, often having more power than the tribe's leaders.* In the diffused light of a new day, Parker finally fell asleep.

The next morning, more tapes and assorted data were delivered to his office. Parker impetuously asked his graduate assistant to take over his remaining classes and then headed for the security of his cottage. Once home, it took a few moments to become accustomed to the grainy, black and white broadcast, but soon he was comfortable with the videos and happy that most of the irrelevant material had been edited out. Perhaps a zoologist would be interested in the hours of movement through the mountainous terrain, the foraging and resting, but Parker found it useless.

The first tape showed Grendel watching two men in an open area about a hundred yards beneath him. The bigger man had a gun, the smaller man a bottle or flask that he often sipped and then shared with his friend. Some time passed before a young doe wandered into the clearing. Slowly the rifleman carefully aimed and then squeezed the trigger. The doe, hit in the rear, partially dropped to the ground, but her front legs continued to churn the rocky soil, attempting to flee. The smaller man patted his larger friend on the back and handed him the bottle. He took a long swig, then, he took another shot, hitting the defenseless animal high on her right foreleg. He laughed, apparently enjoying the animal's suffering. After a few more minutes, the smaller fellow took the gun and fired several shots, missing completely. Both men laughed loudly as if this were the funniest thing that they had ever seen. Each took a few more swigs before the larger man, still sniggering, took aim and blasted off the doe's nose. This prompted more laughter and humorous antics. The animal, thrashing about in a pool of its own blood, was helpless. Another shot rang out, taking away an ear. Again, more laughter. Parker was repulsed. Grendel's vital signs revealed his distress. The smaller man took the gun and began to shoot wildly, the bullets ricocheting off of trees and rocks. He fell to the ground intoxicated, still laughing hysterically. Grendel managed to scamper out of range.

*How long did it take for the doe to bleed to death? Did they let her rot, or did they use the meat?* Because they were both drunk, Parker doubted their ability to recover the carcass so it could be properly butchered. *Man, the insane animal,* he thought, trying to regain his composure. *What kind of perverted creature could enjoy the anguish of a maimed and dying animal? Where did that twisted side of man come from? Was it God-given? Or did lower levels of consciousness within man fester in utterly dark realities, the diminished rooms of damaged psyches? Why were so many people so totally insensitive to suffering, quite often even to their own?* Whatever the answer, the behavior of these two men was both unnatural and ungodly.

There was an abrupt transition. As the tape continued, Grendel watched the priest, Father John, lead a pilgrimage to a holy shrine high in the mountains. It must have been in the dead of summer for the heat seemed oppressive. A cluster of old ladies, all dressed in black, were slowly dragging themselves up the mountain, singing over and over again a chant about sin and salvation. Some were too tired to praise God, while others steadfastly voiced words that as the altitude increased, were becoming more difficult to form. Grendel, a dark shadow in the deep woods, apparently scrambled along with them, higher and higher, hiding behind rocks and trees. Occasionally Parker heard what sounded like a very soft singing, a gentle humming, like a lullaby. *Did Grendel have a simple language, if not words, certain sounds that he associated with feelings or moods, like one might hear from a three or four-year-old when deeply engaged in the moment? Was he imitating the humans that he was following? Was this evidence that Grendel could communicate on some very rudimentary level?*

Bronsky had provided a small packet of interesting research on the origin of primate verbal communication. She thought that there was a strong possibility that Grendel was capable of, if not speech, some type of verbal expression. Tests indicated the full development of the hypoglossal canal at the base of Grendel's tongue. He definitely could form fairly sophisticated sounds, if not words. Bronsky claimed that apes and chimps had a multitude of calls, most likely associated with hand movements and facial gestures. Grendel, no doubt, had a voice box and knew how to use it. Parker wondered, *by being primarily a solitary creature, would he have a need for sound? There appeared to be no other of his species with whom to communicate. Did his brain, nevertheless, use symbols to form observations and possibly basic thoughts? Could he, for instance, associate water with not only thirst but perhaps cleansing? Could he connect darkness with a feeling of comfort that comes from being invisible to mankind?*

Bronsky's notes also referred to neuroplasticity, a new concept that was gaining favor among psychologists and brain researchers alike. Until the end of the twentieth century, she explained, the primate brain was thought to be "hard wired" with each region fixed at birth to perform certain functions. The way the brain worked was thought to be immutable. The latest evidence, however, suggested that brain matter was in a sense plastic and it had the remarkable ability to reconfigurate itself. The Buddhist idea that the self-discipline of a focused mind could actually **change** the way a person perceived reality, **the way the brain actually worked**, was documented by numerous scientists. Apparently our behaviors were the result of **not** some genetic predetermination, but our thoughts which over time had shaped our brains. For instance, brain scans proved that the mind of a practicing Buddhist looked a lot different from that of an ordinary corporate executive. The center for compassion in the person who regularly meditated demonstrated much greater activity than in the brain of the businessmen. Bronsky wrote, "Certainly Grendel, a highly evolved primate, is likely to have this capacity for change. Over the course of his life, through the repetition of the same thought patterns, he could develop certain qualities, like empathy or compassion, just as some people do." Parker was fascinated by these new findings.

*What was Grendel thinking? How was he creating meaning?* Parker continuously asked himself these questions. The observation of the pilgrimage clearly proved that he had a curious nature, but following them all the way to the shrine at the top of the mountain suggested a deeper interest. Perhaps Grendel was forming rudimentary questions about the meaning of what he was witnessing. *What was Grendel's mind doing with its perceptions?* He must have wondered what these frail, old, two legged creatures, clearly not well suited for this activity, were doing. Aging animals dressed mostly in black and exercising during the hottest hours of a scorching day probably appeared as out of place to Grendel as they did to Parker. Why would Father John be lugging, at times dragging, a huge wooden cross up the mountain? Parker had seen animals view human behavior with both discomfort as well as disbelief. As a child he remembered Bonkers, their family dog, hiding under his bed when his mom suddenly screamed about a rip in Parker's T-shirt or dirt on his shoes. Higher mammals were even more inquisitive.

Since the DNA tests proved that Grendel was a close relative to man, with a much higher mental capacity than apes or chimpanzees, the idea that Grendel could create more sophisticated meaning seemed consistent with Parker's vision of him. At this point, the complexity of his thought system could only be debated. Parker's hunch, however, was that this was an animal with the thinking capacity of at least Stone Age man. That, in part, might account for his ability to elude discovery. Maybe no skeletal remains had been found because he buried his dead; perhaps no feces were discovered because he covered his waste. If Grendel's evolution were at the beginning point of human meaning making, by studying Grendel, mankind could have a firsthand look at the earliest stages of his own development. This possibility was the driving force behind Parker's excitement.

At the shrine Grendel carefully situated himself far enough away to go undetected, but with an unobstructed view of the proceedings. The cross that Father John carried up the mountain was in the foreground, dug into the earth and pointing to a crystal blue sky. The pilgrims sat on crude wooden benches. Their faces were drenched in perspiration and their eyes were focused on Father John's every movement. They were tired and thankful, grateful that their long trudge upward was behind them. The heat, however, made the remarkable beauty of the landscape far less enjoyable.

Soon, the Father put on a special headpiece and began a series of rituals that required the parishioners to constantly stand and sit, stand and sit, reciting a series of phrases that Parker interpreted as confessions of their sinful nature. The footage at this point was tilted, as if Grendel dropped his head to the side, perhaps confused by the spectacle that he was viewing. Even for Parker, a professor of religion, the service appeared incredibly surreal, as if it were happening hundreds of years ago near a remote village that time left behind. On a beautiful yet insufferably hot afternoon, these humans chose to distance themselves from the moment and pay homage, Parker thought, to some ancient creed. *Could Grendel form a similar perception?* If animals with less developed brains could interpret human behavior, such as flight being a sign of weakness, certainly Grendel was in the process of ascribing some significance to these actions. Unintendedly, the film cast humans in a rather humorous light. Father John's face seemed squashed under the golden plumage of his gigantic hat. As he continually pointed to the sky and shook his fist, he appeared more like a comedian than a man of the cloth. *What could his antics mean to Grendel?*

Before Parker could construct any deeper meaning, tragedy struck. One of the old worshipers fell heavily to the ground as if violently knocked down by an unseen hand from above. Quickly the dark shrouded pilgrims formed a circle around her, several were crying and one was screaming fitfully; others were praying. Patiently Father John again raised his hands to the heavens, looked up and in a calm, confident voice uttered some words that Parker thought might be Latin.

To Grendel, the act of addressing nature, like the blazing sun or the pure blue sky, might seem foolish. *Was this transferring of an animate energy to inanimate objects in nature the beginning of formal religion, the exact point where humanity departed from the animal kingdom by seeking an inexplicable symbolic connection to something far greater than itself? Could witnessing this act by Father John awaken the same potential in Grendel?* The monitors on the other side of the screen indicated a sharp shift in his vital signs. Grendel apparently felt something, but what he felt could only be speculated. *Did the suffering of other living beings, bipeds, standing erect like himself, cause him anguish?*

Early man certainly practiced some rudimentary rituals. *Did Grendel understand more than anyone realized? Were fear of the unknown and compassion for those experiencing pain the building blocks of the earliest religions?* An accurate analysis of Grendel's response could go a long way to further man's understanding of how religion emerged on this planet. Of course, fundamentalists might be uncomfortable with the findings. Clearly, fear played a large part in both the Old Testament and the Koran. However, the suggestion that man's anguish over the present as well as an uncontrollable future was instrumental in his need to create a Godly father figure, an observation that some would claim as self evident, would not be warmly embraced by traditional thought.

Bronsky's notes on the relationship of the right brain to the development of the divine mind were very insightful. Her position was that early man had little or no ego; therefore his experience of reality was more vital than his thinking about it. There was far less "stuff" to interfere with the moment. *Parker questioned how Grendel's experiences changed his understanding of reality? Was his consciousness perpetually shifting from images to bodily sensations to vague states of emotion? Did his poorly developed sense of self allow him to be both more intuitive and more likely to embrace life through changing states of consciousness, rather than rational thought with its incessant stream of never-ending words?* Obviously Grendel was not influenced by any form of society. His freedom, Bronsky theorized, allowed him to strongly engage the physical world while simultaneously intensely experiencing a rich inner reality. His mind was not a jumble of thoughts but, more likely, a center of intuitive awareness that facilitated a unique understanding of his surroundings. Parker wondered *through meditation, an act witnessed on an earlier tape, did Grendel enter hypnologic states that allowed him to transcend both time and space, momentarily putting him in touch with the unifying energy at the center of the universe?*

Bronsky conjectured that Grendel's brain was better suited than modern man's for merging with a sense of oneness, what one might call an encounter with the sacred or the oceanic condition. Egolessness is what the eastern traditions called it. Grendel was the Taoist's uncarved block, a simple, natural, largely uncorrupted vitality, the very condition necessary to fully embrace divinity. He was not burdened with societal instructions and the games that people were forced to play. He was not forced to wear the foolish masks and to endure the required posturing that stripped man of his birthright to directly encounter the Ultimate Power. *Could he be empowered in ways that man was not?*

She then supported her position by referring to the firewalkers of the South Pacific. After a few days of intense preparation, producing a deep spiritual transformation, they joyfully danced across burning coals as hot as 400 degrees Fahrenheit. They were in no hurry to cross the ashes and clearly felt no pain. Better educated people, however, had much less success. Because westerners thought walking on scorching hot coals without burning one's feet was literally impossible, they could only witness, but never directly experience the ecstasy of the full ceremony. In this case, apparently the more "evolved," mentally developed person was denied access to a desirable state of being. Parker, too, was beginning to consider a startling possibility, one that Bronsky, perhaps, was implying: Grendel was not just different from modern humanity, **but** in some ways superior.

The right brain appeared to have childlike qualities that many world religions, especially Christianity, seemed to glorify. Parker recalled Christ's instructions that to enter the kingdom of heaven, one must believe as a child. Apparently the sophistication of a well-educated brain was a barrier that usually hindered real transcendence. The Bible often warned man about the pride of knowing. On some level all the information in the world could never lead man to the truth that his left brain required. A simple, uncluttered mind, largely free from social constraints, was the ticket to the grace of higher states of being, and perhaps, to actually encountering divinity.

The Biblical quotation, "Be still and know that I am God," also seemed pertinent to Parker. The condition of silence was necessary if one were to be able to feel the presence of the sacred. He often cautioned his students that our culture was totally preoccupied with mindless distractions. We were all in a hurry to accomplish certain goals that had little if any bearing on the big picture, on a quality existence. Most of the media's constant drivel moved the individual's sense of self to peripheral and quite trivial activities. The vast majority of people, he believed, were so busy with these largely meaningless experiences, like excessive material consumption and perpetual pleasure seeking, that they could never return, or only momentarily, to one's birthright of natural innocence. *Had Grendel evolved with an uncontaminated mind, reducing life to a minimum of worldly details while living in a childlike state of grace, a Garden of Eden of sorts? If so, was that not a state of being that was far more capable of knowing and uniting with the sacred than the overburdened souls produced by the restrictive dogma of most world religions? Was not deep within Grendel the true light through which Christ, Allah and the Buddha encouraged man to live?*

# CHAPTER SIX: GROWING CONFLICT

The phone's ring, a harsh return to reality, forced Parker to leave his happy musings and confront life in the present. He was shocked to hear the voice of the president of the university. "Eric, Paul Knight. I'm sorry to bother you at home, but we have a bit of a problem. I hate to burden you with this, but the university has received a number of calls from people... I promised that they would remain anonymous... regarding the Grendel Project. One of them, the person didn't identify himself...I hate to say was quite disturbing, suggesting that the university might come under fire if we did not repudiate 'this missing link nonsense.'"

"I'm sorry to hear that. Were they at all specific," Parker inquired?

"That's the problem, Eric; I'm not sure whether the fire they referred to was suggesting some type of physical destruction, or worse, perhaps a mass shooting or something of that nature. Maybe it was merely a colloquialism for 'hot water,' I hope that is it, but I feel that I can't take a chance in the volatile climate of today's university life... particularly with the recent shooting at Virginia Tech."

"I understand."

"Also, some important benefactors, again I'm afraid that I can't mention their names, have contacted me, asking that the university drop our sponsoring of the Grendel Project."

"What was their justification...I mean this is vital research. What's the problem?"

"Everything is a bit vague, but they used phrases like 'offensive to one's convictions' or 'damaging to the school's prestige.' Eric, you know philosophically I believe in investigation and the search for truth, but I feel the most sensible course of action is to downplay this whole thing for the present, but not entirely eliminate the future of the research."

"Paul, I'm willing to ride it out," Parker said with a deep sigh. "I've already paid a big price...with threatening phone calls, letters, emails, and you know about the assault, but this research is an opportunity that probably will never come my way again."

"Eric, I think things are getting out of hand. I hate to do this, but for the good of everybody, I think we need to take a break from this program. Down the road, six months to a year, maybe we can return to it, but for now, let's consider it closed," Knight said, his voice suddenly revealing an irritable tone.

"Paul, we can't overlook the incredible significance of the program. It could truly enhance the university's reputation in the eyes of the academic world. We are on the verge of a revolutionary new understanding of not just another species, but of mankind's earliest stage of development."

"Eric, either you walk away from this right now or else I'm going to request you take an emergency sabbatical leave of absence, and that would look awfully peculiar in the middle of a semester," he said, almost shouting into the phone.

Parker paused, sighed again and then said flatly, "Well, I guess I have no choice. I'll drop it for the time being."

"Thanks Eric," Knight said, his words feigning warmth that was not sincere. "I know that this is hard on you, but it is best for everyone involved. Let me know if you have any more problems. We'll keep in touch. My best to Kathy. Goodbye."

Parker was shocked! He was being shut down, asked to forfeit the dream of a lifetime. The idea of making minimal, or even worse, no progress for the next few months or even a year, did not sit well with the professor of religion who would permit nothing to obstruct his pursuit of a breakthrough that could dramatically change not only his life but add new perspective to the early years of man's evolution.

Defiantly, he returned to the video tape labeled SUMMER. Within a few seconds Knight's demanding voice muted as Parker became mesmerized by the action. Grendel was quietly drinking from a pond beneath a series of cascades, the cold, mountain water running down his chin and over his body. Suddenly he plunged headfirst into the pool, like a teenager at a swim club. For a few minutes he was a child, kicking and splashing about, submerging himself and holding his breath, sometimes for as long as a minute. Slowly he settled into the experience, floating serenely on his back, watching the sun, like dozens of sparkling diamonds, glitter through a canopy of large leaves. On the other side of the screen his vital signs suggested that he might actually be moving into a trance or, perhaps, a state of deep meditation. Parker thought that if he could see Grendel's face, it would reflect an expression of pure contentment. No doubt he was focused deep within himself, on what theologians might call his spirit, and judging from his alpha brain waves, he was in a state of pure joy. His breathing was slow and rhythmic. This continued for over five minutes. He could hear Grendel's long arms gracefully paddling, gently keeping him afloat. For the moment Grendel was full of a glorious energy that was imbued with pure light.

As the camera slowly wavered up and down, recording the lush foliage overhead, for an instant a person's face seemed to appear and just as quickly disappear. Startled, Parker pressed pause, and then attempted to enlarge the image. From what he could observe, it was the bearded face of an older man, peering down from the rocks above. At that same moment, the alarm signals triggered in Grendel's brain and he was in a full fledged panic mode, scrambling at top speed down the rocky stream bed. The hunters knew that they could never catch up to him, but nonetheless, they started to chase him down the mountain. *What was their plan? Was there a trap or cage awaiting him below?* Suddenly Grendel veered to the right, scampered over several large boulders into a cave and the safety of semi-darkness. For the time being he had eluded his pursuers. His vital signs slowly returned to normal, but before the sequence ended, the monitors spiked again. Grendel's whole body tensed as a blood chilling scream echoed off the surrounding rocks. *What happened?* Parker concluded that if something unfortunate befell Grendel, it would certainly be captured on video. *Grendel must be safe, but what was that sound?*

Curious to know more, Parker began to sort through the anecdotal information that accompanied the tapes. One of the articles provided was an interesting news story from a small community deep within the mountains. The date corresponded to the approximate time frame of the tape. It presented an account of how The Giant from the Darkness killed a local hero. Apparently the man's body was found horribly mangled at the base of a ravine. Two eyewitnesses claimed that The Giant from the Darkness cowardly attacked him from behind, pushing him off the cliff. They said that before they could get a clear shot at him, the creature disappeared behind some massive boulders.

The news article suggested that encounters with the Giant from the Darkness were not totally uncommon. This, however, was the first murder attributed to the huge creature. *Ridiculous,* Parker thought. *Grendel's activities were recorded. When the scream occurred, Grendel was hiding in a cave. Most likely the man slipped or, perhaps even worse, his friends "assisted" him off the ridge, but Grendel was clearly innocent.*

Now, even more people might be pursuing him which could interfere with the team's research and turn the Grendel Project into something like the UFO craze of the 50's. As Parker further reflected on the news article, it occurred to him that the world press avoided this type of story, considering it "questionable" at best and "beyond credibility" at worst. But how many villages in the mountains of Eastern Europe, or in the western part of the United States for that matter, have had a citing of a large, hairy primate, if not direct encounters? *Unless people take this seriously, we're never going to get the full meaning of what's here,* he thought.

A second essay, an opinion piece from the editors, was stapled to the news story. The title, loosely translated, was "Monster Madness, or Stories of Hallucinations?" The authors listed numerous "encounters" of a strange creature during the past few decades and suggested that there was one simple explanation: when a person sees something that they don't recognize or don't understand, many minds categorize the unexplained phenomena as "unnatural" or "out of this world." The rationale goes something like this: a person sees the rear of a black cow in a thicket. Since cows are always on pastures, the mind doesn't comprehend that it is a cow, interpreting their perception instead as something totally bizarre or from another world. The "strange" creature suddenly becomes a "monster," the only explanation that makes sense.

The editors further suggested that because our lives are filled with repetition and routine, our unconscious mind actually enjoys sabotaging reality, spicing it up with scary ideas. This makes our days more exciting and our rather banal existences more bearable. Having a strange creature in the area might be fun. Since no indisputable fossil evidence or decomposed body had ever been produced, the authors were certain that "monsters" were figments of man's active imagination. If mankind can "see" angels and ghosts, then "monsters" are not much of a stretch.

*They'll be amazed,* flashed across Parker's mind as he envisioned the two editors viewing the tapes for the first time. His thoughts, however, were interrupted by the telephone's ring.

"Eric, Steve Herder. I have some disturbing information from a pretty solid source that Islamic extremist have put an elaborate plot into action to take you out. I don't know exactly when, but soon." His words sounded rushed, as if he could not get them out quickly enough.

"What did you hear," Parker asked, alarmed.

"The internet chatter has increased greatly. It's really serious. Apparently a Syrian newspaper got hold of a story about you and Grendel in the American press. Its publication stirred up a great deal of outrage. That's all I know. Listen, if you have some emergency plans, now is the time to put them in play."

"You must know something more! Are they going to hit my home, my family, the university… there has to be something, some details," Parker asked, pleading for information.

"I've got nothing more. Listen, the best I can do…if you need a place to stay, you can use my house for a couple of days."

"Thanks, Steven, but I'll handle things…I'll be Okay," he said unconvincingly, his eyes staring dejectedly out of the window.

That evening was extremely painful, as if time were slowly dragging its bleeding carcass over events too tormenting to endure. While Matthew and Mark were getting a few hours of sleep, he agonizingly shared with Kathy the substance of Herder's phone call. "I don't think that we can wait any longer," he said, almost in a whisper, as if he couldn't bear to acknowledge his own words. "You're going to have to take a temporary leave of absence from your teaching position and take the kids to your sister's. I think she is far enough away. We don't have any other choice." He swallowed hard and then added, "I still don't believe that it has come to this," while looking down at the floor, shaking his head.

"Eric, you have nobody to blame but yourself," she said angrily. "I begged you to give up the project, but no, you knew better. Look where you and your big dreams have got us! You let it happen; now we all have to pay," she cried, the tears rolling from her eyes.

For a while they worked in silence, packing their cloths into suitcases and clearing the kid's bureau drawers into paper bags. Months ago they both agreed that the children's welfare was their top priority. They knew from now on that there could be little if any contact with each other. Emails and phone calls could be traced. For the short term, at least, they both were going to be totally on their own, for better or worse, struggling in their own separate realities. While most of their necessities were being shoved haphazardly into their SUV, Kathy continued to vent her rage. "How could science, no matter how revolutionary, be more important than our family? Eric, this is off the God damn charts! I'm giving up my job, my friends, my house, my whole damn life for your fucking Grendel."

"Kathy, it will work out, I promise you."

"Don't make another promise that you can't keep," she screamed, repressing her hatred. "You really fooled me. I believed your universal compassion and unconditional love bullshit. Eric, you're the worse kind of a hypocrite; you don't understand that your life is…the worse kind of dishonesty. You talk about striving for the highest while your family is shattered. All of your big ideas don't mean a damn thing if you're dead and we have to fear for our lives. You just don't get it!"

"Kathy, I love you. We'll get through this."

"I don't see how," she said coldly. "Love is not a pretty little idea that you lock away in your head somewhere. Sometimes I wonder if you have any emotions left in you at all, with all of your meditation and Buddhism bullshit.

You seem so distant," she said, her eyes glaring wildly at him. "So don't talk to me about love! If you love anything, it's your Grendel Project, your books and your TV shows…it certainly isn't us. Eric, I'm sorry to say this," she said as her lips trembled uncontrollably. "You need to stay away, I really mean it, until everything has completely settled down, until everything is *absolutely* safe…and even then, I can't make any promises about our future."

"The last thing I want, believe me, is for anything to happen to you or the boys!"

As he embraced his family for what might be the last time, he felt the numbness of disbelief. *This really wasn't happening, was it?* He repressed his tears. The boys were too young to understand, but they knew something was terribly wrong. Matthew held onto his teddy bear while Mark clutched his favorite blanket. He assured them, "Daddy loves you, but we all have to go away for a little while. It will all work out," as he buckled them into their safety seats. He tried to smile. Then he hugged and kissed each of them.

Their goodbye was hurried, both of their minds already thinking about the challenges in their immediate futures. It was already dark when Kathy pulled out of the driveway. Their SUV was filled with the boy's clothing and toys, as well as her personal items and most prized possessions. What had been a reasonably happy marriage, at least from Parker's point of view, suddenly seemed precariously stretched out of shape, like an over inflated tire that might explode if it hit many more bumps. When Kathy was safely on the road, he shut off all of the lights and tried to make his house appear unoccupied.

In his vacated bedroom he sat on the floor, his back against the wall, and lit a small candle that danced frantically in the erratic breeze from a slightly opened window. In the semi darkness of his broken world he assessed his bleak future. Kathy's words, "you don't know what love is," rumbled around his head like distant thunder. He was painfully aware of their differences!

For her, love was all about emotion, strong attractions for something or someone. It was something that you felt in your bones. She would never understand that feelings were unreliable and changing from day to day…that emotion terribly distorted the truth. The notion behind the Hindu greeting, "namaste," best captured his sense of love. It was simple: "may the beautiful, sacred energy within you be one with the beautiful, sacred energy within me." Love was opening the heart and honestly embracing your beloved, as if she were part of you. People who genuinely shared their beings avoided the senseless rollercoaster ride of fluctuating emotions. He wondered, *why couldn't she accept his pursuit of truth? Why couldn't she embrace his being as he was, not as she wanted him to be?*

Marriage, as he thought of it, was between two centered beings who openly shared their journeys, both the joys and heartaches. And in that way love was an unshakable union, bonded together not by what a person felt but who a person was. Suddenly, a gust of wind extinguished the candle. The blackness quickly brought him back to his plight. He hoped that he had a little more time to create a new identity and a corresponding shift to a new reality. As he sat in total darkness, he had absolutely no idea of how he was going to accomplish that feat.

The next morning, after a mostly sleepless night, he returned to the university to attend to a few last minute concerns and teach his final class. He said nothing about his future while trying to pretend that everything was normal. "The idea of an avatar, the notion of a truly enlightened person who has directly descended from God, is prevalent in many religions. The word avatar is Hindu in origin and essentially means 'embodiment of the divine.' Hindus, however, don't simply include only the exceptional personages of their religion, but openly embrace figures like Jesus, Zoroaster and the Buddha. Each avatar has come to restore a balance to life on this planet, each having some unique insight into the nature of absolute truth. By honoring the avatar's pure, gracious energy, which reflects a highly evolved level of consciousness, mankind can actually witness a blissfully divine manifestation of the Ultimate Power," he paused a moment, allowing this idealized view of man to sink into his consciousness.

"Freed from the bonds of time and space, the avatar's actions are direct expressions of the most sacred. Mother Theresa, Gandhi and the Dalai Lama are some twentieth century avatars, viewed as most holy by their followers. Many world traditions, mostly eastern, believe that each person has the potential within them to achieve this elevated state, what I refer to as 'avatar consciousness.' In fact, they might claim that this lofty condition is man's natural state, the true intention of our lives."

He fumbled around looking for something in his briefcase. He found an old paperback book and holding it up, said to the class, "Herman Hesse's novel, *Siddhartha,* is a beautiful rendering of the divine as it is embodied in a physical form. When you get a chance, I suggest that you read it. It explores the Buddha's struggle with worldly temptations, and in the end, his ability to transcend all of the pitfalls of material trappings. Because of our attachment to our egos, to our cultural identity, most of us never embark on this most holy of journeys. Of course, our society grossly devalues it. Avatar consciousness, however, is the ultimate state of being to which we all should aspire. It is, if you will, what exists within each of us when our social posturing is stripped away and we live freely and spontaneously as we were formed by our Creator.

Simply put, the more attached we are to the world, the further removed we are from divinity," he said with a sigh.

As Parker concluded this final thought, he gently touched the jagged scar on his forehead which was still sensitive. He was unconsciously troubled by the imposition of the worldly on his immediate future. Rather than cultivate a discussion, his usual procedure at this time, he abruptly thanked the class for their attention. He told them that they were free to leave as he absentmindedly stuffed the somewhat battered book and his notes into his briefcase. He departed rapidly, walking off into the unknown without any conscious fear of the future. Often he had advocated to his classes that true living was no more than "surfing on the crest of the eternal now." It was his turn to live it. He did not look back or even wonder about the reactions of his students. For the moment at least, he had escaped all of the responsibilities that had given his life definition; shortly, he would be completely free to create himself anew.

While feeling a sense of both total dislocation and incredible exhilaration, Parker walked to his car and then placed his laptop computer and several cardboard boxes of research on the back seat. As soon as he stepped off campus, he would be nobody, a person with no particular title, existing in the moment with no neatly carved out place in society. He was free of his culture's insidious institutions and their binding set of expectations. His only obligation was to stay alive and attempt to ride out the turbulence of the unexpected storm.

Parker felt a strange parallel in his life with the lesson that he just taught. The avatar experienced numerous conflicts with the secular world, often was condemned, yet always managed to maintain his natural, pure spirit. He remembered a college professor once stating that the highest, most evolved person in a group, the one who should be most revered, was not the person with the most power or the most refinement or the most education, as one might expect, but the one with the most cultivated soul.

Suddenly he wondered if the world had everything in reverse. *Was the pure, flowing energy of higher mammals more elevated than the selfish agendas of most well educated, highly successful humans? Grendel,* Parker thought, *might be right at the center of the crossroads, a highly evolved primate, with a consciousness or spirit, yet without the destructive messages that emerged from an overdeveloped ego immersed in a contaminated society. That idea would deeply anger most of mankind: Grendel, the undiscovered avatar.*

With the car windows wide open and the autumn air blowing in his face, he felt his life had become a kaleidoscope of sorts, a series of constantly changing moods and panoramas, shifting from anxious thoughts to happy possibilities and back again. Soon he would be home, fill a satchel with the

Grendel tapes and the related information, pick up his suitcase and then off to Herder's for the night. For the moment his mind was on autopilot, enjoying the beauty of the changing foliage as his car effortlessly traveled the route it had most days over the past seven years.

When he slowed for a stop sign, a car suddenly pulled out of a nearby driveway and raced to overtake him. In no time, it was stopped next to him, blocking the oncoming traffic, but there was none. In fact, there was no one in sight. Instinctively, Parker looked to his left. A man, his face covered by an incredibly realistic looking gorilla mask, shouted, "Parker, you foool," and then from the passenger seat he raised and pointed a handgun at him. Parker immediately dove onto the floor as several bullets shattered the front window and one pierced the car door. The assailant quickly pulled away, his tires squealing on the black asphalt.

Parker was covered with fragments of thick glass which resulted in numerous minor abrasions. Fortunately he was not seriously injured. Stunned, after what seemed like an eternity, he slowly moved, opened the passenger door and crawled onto the sidewalk, the heavy shards of glass falling from him like hundreds of dead, translucent insects dropping without struggle onto the rough concrete. Dazed, unable to think or react, he sat on the ground and covered his head in his arms. Then he sobbed the slow, muted cry of one who mentally understood, but was emotionally devastated, painfully suffering from an invisible wound far deeper than any surgeon's incision. He felt completely shattered, as if his brain were being attacked by hundreds of voracious insects, each a sharp shard of dirty glass frantically drinking his blood and eating away at his sanity. He struggled to retain consciousness as his world went black from time to time.

Several drivers passed slowly, one called, "Are you alright?" No one stopped! Instead they dialed 9-1-1 and reported what appeared to be a one car accident. The police arrived and after making certain that he was not seriously injured, began their investigation. Hoping to get information regarding the bullets and ultimately the gun used in the shooting, they impounded the car. At the police station, Parker calmly described, several times, in an extremely detached voice, everything that he could remember. He could provide no license plate number, no car model, although he thought it was a dark color, not even the race of the shooter. All he could remember was that a man, with a gorilla mask and probably pretending to be Grendel for ironic effect, called him a fool and then opened fire…three, maybe four shots and lots of splintered glass. That was it!

After several hours of fruitless questioning and a review of the details of the earlier incidents, the threatening phone calls as well as the email chatter, Herder arrived and they drove back to Parker's cottage. They cautiously entered and Parker quickly retrieved the tapes and his suitcase of already packed clothing. If he were thankful for something, it was that Kathy and the boys were protected and would know nothing of this ugly series of events.

Once in the safety of Herder's house, too overwhelmed to sleep, Parker renewed his investigation, desperate to find the truth. The Grendel Project pressed heavily on his bewildered consciousness. He had to trace it to its conclusion. In Herder's den he placed the next tape in the video deck, rarely used now, and started to analyze the segment. He was happy to return to Grendel's black and white reality which, far removed from his plight, offered a strange sense of security.

The camera angle indicated that Grendel was observing from a high vantage point, his usual perch, where he could see what appeared to be a funeral procession. Most of the community appeared to attend, honoring what the newspaper called a "fallen Hero." Six of the strongest men slowly carried the victim along the main thoroughfare in a specially designed box that was decorated with golden ornaments, a dozen different swords and shields. Parker sensed a strong militaristic energy among the mourners, many standing at attention and saluting as the coffin passed by. Aggression seemed to be something that most of the men understood and accepted. *Perhaps they enjoyed it*, Parker thought. On their lapels many wore an insignia, what looked like a giant clenched fist with a crucifix dangling from it, a strange union of Christianity and warfare. *These "God fearing" people*, Parker thought, *seemed awfully comfortable with force.* By comparison, Grendel appeared serene, if not totally harmless.

Father John wore a magnificent red robe, adorned with intricate designs of golden crosses and opulent crowns. On his glorious headpiece there was the image of a man being crucified. When he rose majestically and held his hands toward the heavens, his audience immediately hushed. Grendel studied their faces; some were crying, but most seemed terribly depressed. The priest then eulogized the "hero," calling him an exceptional man who died doing the work of God. "In attempting to eradicate evil from our small community, from the mountains that we all love," Father John asserted, "he met a brutal as well as undeserved demise. For this, God will embrace him with open arms and he will take his special place in heaven, knowing that he was a good servant in the fight against the dark, satanic forces of the universe."

Parker, perhaps more sensitive to life's paradoxes than usual, was stunned by the contradictions. A shy, nocturnal primate, seeking to live peacefully in isolation, was labeled "evil" and according to Father John should be "eradicated." It was clear to Parker, any living thing that man doesn't understand will be invariably labeled a "monster," and according to many of our religions, something not only to despise but to destroy. Parker remembered the Bible's treatment of the serpent. In fact, according to the Old Testament, God created all animals on Earth to be "subjugated" by man, suggesting that anything that is not human has no soul and no true "right" to life, at least not the same kind of right to life as man. *How could the sacred,* Parker wondered, *become associated with ideals that represented such a poorly evolved consciousness and one that seemed not just to tolerate but actually encourage violence?*

Even though Grendel could not fully comprehend the spectacle that he was witnessing with all of their supposed metaphysical meanings, he was fascinated by the bright colors and elaborate rituals. Grendel spent most of the time with his eyes fixed on Father John, as if he, Grendel, were one of the parishioners. For minutes at a time he stared at the dead man, not in the casket but on the priest's headpiece. Given all of the possible symbols in the universe, Parker wondered why mankind would choose as most holy an image of such a gruesome death. Of course he knew the doctrines of the church and the assertion of resurrection and life everlasting Just the same, it was barbaric to associate God's mercy and unconditional love with a man brutally spiked to a wooden cross and helplessly bleeding to death. Somehow it made God seem comically impotent. Where was His omnipotence when He needed it? A sense of futility surged through his being. Like Grendel, he was being hunted by "God-fearing people" whose stated intention was only to strike down evil, which to them was anything that differed from their fundamentalist understanding of their proclaimed holy scriptures. They would never be able to grasp a new way of thinking about what might constitute a genuinely religious life.

Parker's attention was drawn to a packet labeled "anecdotal." It contained the translation of dozens of journals that were initially started in 1996, within a year of the Grendel tapings. They were written by an anonymous female, beginning when she was approximately eight or nine years of age and continuing into her late teens. Her foster mother had collected them over many years. Then somehow she contacted Bronsky, hoping that her world renowned psychological expertise might be able to provide insight on her child's illness. The psychologist, however, was unable to provide the hoped for cure.

Parker was immediately struck by the intense rage conveyed by the girl's handwriting. There was nothing flowing or pretty about her script. Each letter was impaled deep into the page as if she were stabbing a hated assailant. Nightly, the young girl attempted to capture in her diary her troubling dreams, perhaps better described as reoccurring nightmares. In the darkness she was always in the midst of unimaginable carnage. People that she knew and loved were being assaulted and slaughtered in cold blood. Their necks, she wrote, often were severed as if they were barnyard animals being prepared for the market.

In her nightmare she was frantically fleeing, running for her life. Yet as fast as she could run, no matter how hard she tried, she was never able to distance herself from the savage butchery. At any moment she expected to die, struck from behind, yet she was afraid to turn around. Somehow she knew that what she was trying to escape from was more horrible than words could ever describe. At times she wrote with a red ink that when she pressed too hard, ran wildly onto the page like blotches of blood.

Then, dream after dream, out of the smoke and chaos a large black face appeared, with sad eyes and a kind expression. He was a massive creature with wide cheekbones, a sloping forehead and extremely hairy. Clearly he was not a human being. Then, night after night, she felt a strong arm reach out and lift her above the mayhem, almost as if some superhuman creature, a god perhaps, was miraculously carrying her to safety, sparing her life. When she looked at her savior, between his eyes he had a beautiful, shiny star which glittered in the sunlight. The sparkling diamond, in her mind a symbol of his holiness, seemed to mesmerize her, giving her a sense of serenity in the surrounding madness.

Each account, Parker noticed, was worded almost exactly the same: a huge, powerful creature protected her, risking his life while successfully evading her pursuers who for some reason she never described. The one difference that most interested him was that she started by calling him an "animal" and later thought of him as "a special friend." And finally, although her "savior" never said a word, he somehow communicated, maybe with a gentle humming, a feeling that she was safe, that she had been spared from certain death. All she could remember was floating over large boulders, flying almost, as if she were a feather being wafted up into a brilliant, sunlit sky, almost as if being mercifully lifted into heaven.

The sequence of events never changed, yet over the years of writing, her penmanship improved remarkably. She came to truly admire "her savior", calling him "lord" and describing him as "Christ like." At times, when her present life became too perplexing, her entries beseeched him for help or, if possible, direct intervention. She begged him to return and mysteriously resurrect her, again lifting her out of her darkness to a heaven of sorts. Although she had no name for him, it was clear that she loved him as if he were her own father, or maybe as if he were a Universal Father.

Reading more deeply into her personal history, Parker learned that both of her parents "disappeared" and that she was found mute, on the steps of the main church in Foca, a nearby town. No one knew how she got there. The doctors who evaluated her noted that she suffered from a type of selective amnesia that had erased everything from her conscious awareness. She could remember absolutely nothing, including her name or where she was from, not even her discovery on the church steps.

Parker wondered if there could be a connection between the young, Muslim artist, KA, who repeatedly drew images of Grendel and this anonymous girl. Both had lived in the same general area and both had experienced something seemingly "out of this world." Each child described their personal trauma in their own way; nevertheless, there was a similar reference point that suggested that this was an identical event. *Could this be a simple coincidence? Unlikely! Was Grendel somehow involved? Most definitely!* Suddenly he was absolutely certain that Grendel was the source of both children's preoccupations. His third eye, the micro chip camera neatly stitched into his forehead, could be the only reasonable explanation, but Parker wondered, would he ever learn exactly what happened? It seemed unlikely.

There were several more enclosures from Bronsky in the packet, but nothing from Price who, strangely enough, seemed to have lost interest in the project. Did the professor of anthropology find that he had nothing more to contribute, or like Parker, was he the subject of death threats and other forms of harassment? *Is all this torment worth it*, Parker asked himself. Then he thought of the persecution of those who went before him, the imprisonment of Galileo and the burning to death of Giordano Bruno. *This is a historic moment and it calls for unprecedented commitment*, he assured himself, sitting back on the couch, staring at the now blank TV screen. The completed sequence had become a series of black and white dots, gray images that seemed as indistinct as his future. At that moment his life was a gigantic question mark begging for answers. Yet when he left Herder's, *where was he going to go, and more importantly, why? How could he complete the Grendel project and definitively answer the legitimate questions that serious scholars had put forth? He, too, passionately sought answers for the legitimate concerns that other scholars posed. More importantly, could he keep himself and his family protected? Could one person, as he hoped, make a real difference in the twenty-first century? And most importantly, was there any way that he could go to the source and learn the truth for himself?*

# CHAPTER SEVEN: BETRAYAL

He opened his eyes to a cloudy morning and watched through the window thousands of leaves quietly falling from the large oak trees that surrounded Herder's stately home. Parker was mesmerized by their continual movement and for a few moments forgot about his plight. Nature always had a way of soothing his suffering. Then it occurred to him that he was just as dislodged, and in his own way, strangely dancing head first towards an uncertain future, an unnatural winter. For a moment his mind turned to the story of Icarus, his wax wings melting as he flew too close to the sun, the source of ultimate truth. Parker could feel the desperation of the young man's free fall back to a disbelieving earth. He wished that his transition would be quiet and uneventful, more like the descent of the dancing leaves.

Suddenly he was overwhelmed by a deep feeling of despair that filled his inner being with a sense of hopelessness and left him longing for Kathy. During their marriage they were separated very infrequently, only when he was required to deliver a speech at a convention for religion professors or when she tended to her ill father. He was saddened further by the realization that he had no idea when he might see her again. Because they had decided that any type of communication could put their family in danger, they would have to have faith that each other's lives were as they should be.

His mind slipped back to the beginning of their relationship. He had fallen in love with her smile, a buoyant, uplifting energy that could turn one of life's ordinary moments, like eating dinner or folding laundry, into a joyous meadow of beautiful butterflies. He remembered a Sunday morning when they walked hand in hand on a trail by a meandering river. They came upon a field of wildflowers, the colors absolutely spectacular, and her face was all aglow. He wanted to pick a glorious bouquet and take it home as a tribute to their love, but smiling broadly she refused, reminding him that she loved living blossoms. In the beautiful sunlight she danced as if she were a blithe fairy, occasionally blowing him kisses as she seemed to float in paradise. Such was their happiness. He had felt that she was blessed with a special gift, a

God endowed quality that radiated an innocent love for everything living. She was irresistible! She expanded his world in ways that he never dreamt of, and at that moment, their love elevated his spirit as never before. In their early years she was the glorious light that drew him upward, an intoxicating blissfulness that combined the joys of earthly and heavenly union.

In some mysterious manner, he hoped his intimate bond with Kathy would bring him permanently closer to the glorious universal energy that he so passionately sought. Now, he missed her and those moments of serenity more than he ever thought possible. He had taken so much for granted. Sadly, while they both tended to their roles as professionals and parents, they began to slowly drift apart. Several images of their simple, joyful family life floated in and out of his mind like the tumbling leaves. A walk on the beach, roughhousing with the boys, camping out in the mountains, all those happy activities seemed so very far away. Already their marriage was beyond those pleasant snapshots. *Would he ever experience them again? Was his longing for Kathy and the elevation that she once provided a futile desire to return to the fleeting sensual pleasures of their youthful lovemaking?*

*Life was truly absurd, he thought to himself; one minute everything was as it should be, and the next second it seemed, it was all gone, ripped away by the most improbable turn of events.* Nobody can go back in time. As he remembered the teachings of the Buddha, in particular the second noble truth, he confessed to himself, *it's almost impossible not to be attached to those you love and to moments of joy even if they were only infatuation.* Yet, at that moment he could not envision how his existence, now so alienated, as well as his family's life, was going to play out. His hope for a happy reunion was the driving force behind every waking moment. It was all that he had left. He did not believe that God preordained each of our fates, but that we, through our decision making process, determined our futures. He would find a way. He had to! Everything had to be balanced out; compromises had to be made. There had to be a way that his love for Kathy and the boys could survive his passionate quest for clarity and meaning.

Sadly, he remembered the discontentment of the more recent years, and how the joy that once illuminated his world like a June sunrise was reduced to something on par with a five watt bulb. She had shut down, internally building most of the usual comfortable walls that her upscale, suburban society required, and in doing so, creating major obstacles in their marriage. Perhaps he had done the same. Unknowingly, their mutual dreams slowly deteriorated and unconsciously they both began to distrust one another. At that very second, all he could feel was the sting of incredible loss. The summer dawn that had so much promise had become a winter's sunset, cold and foreboding. There was nothing of value left in his life. His once thriving career, if not completely dead, was at best "on hold;" his marriage was suffering in ways that he never imagined possible, and his confidence in mankind was at an all time low.

Ironically, his plight, his damaged life, was a product of his unwavering commitment to his search for truth. *What would have happened if Copernicus decided that it really didn't matter that much that the earth rotated around the sun and not the reverse? But the great scientist,* Parker reminded himself, *waited until after his death before allowing the results of his works to be published. Nevertheless, if the facts that mankind used to construct a clear picture of reality were inaccurate, was he not obligated to correct what was false and to do it now?* The Grendel Project could accomplish just that: establish a new paradigm for understanding who we are and how our sense of divinity was shaped. Nobody should be forced to relinquish their personal quest for meaning, especially one that could reshape man's picture about his destiny, and maybe even save the earth. Sometimes his idealism amazed even him.

Herder had breakfast ready and Parker ate as if he were a teenager again. He had forgotten what it was like to actually feel hungry. "Thanks for the bacon, eggs, toast and honey. And thanks for helping me out with a place to stay. Without you, I might be dead by now," he said trying to make light of a serious matter. Suddenly his eyes were drawn to the kitchen counter and the face on the screen of the small, portable TV. It was a member of the Grendel Project, Aaron Price, appearing on a morning news program. As he turned up the volume, Parker heard the scientist assert, with no uncertainty, that "the Grendel Tapes are a fraud." Parker's head snapped to attention as all of his energy focused on that small set, his universe reduced to an eight by ten inch face.

"I am ashamed to say that I was affiliated with a project that was put together solely to discredit traditional religion and call attention to several professors' own mediocre careers. Let me set the record straight. There is no new evidence of a missing link. The videos that have been supplied to various TV stations-most of you have seen them by now-are nothing more than carefully fabricated footage, taken by various film makers, with a hand held camera. Then the material was craftily edited in a film studio. It is all a big act. I know that most of you never believed this Grendel nonsense to begin with. An expert camera man can make almost anything seem real. And the monitors that you see on the side of the screen were added much later and manipulated to go with the selected images and the interpretations that they wanted you to believe. That part was easy," he said in a hollow voice as if he were reading the words from a monitor. He seemed distracted and distant. *Was his life in danger?*

"We live in a frightening time, when virtually any image can be manufactured and appear as real. Digital photography can not be trusted. For that reason, it can not be used as evidence in a court of law. If more images should show up, and I doubt that they will, don't give any credibility to them." After a pause, he added, "They can only be more of the same fabricated material that you already have witnessed. In recent years, science has received a great deal of criticism for a large number of reasons, some justifiable, some not. The so called Grendel Tapes are a discredit to the millions of dedicated scientists who are working for noble causes, like a cure for cancer or an inexpensive alternative fuel. I am embarrassed that I was associated with this blatant scam. Please accept my apologies." Price seemed uncomfortable, his eyes often avoiding the camera. He was in a hurry to end his presentation.

Several reporters had questions. "The DNA and blood sample, for that matter, all of the objective data, was that bogus also? If so, how did they pull off such a hoax?"

"The group appeared to be reputable scientists, but in fact, they were professors with an agenda. None of the 'data' was confirmed by an independent laboratory...so there really was no reliable corroboration. They made up whatever they wanted and passed it off as scientific fact. Their publicity seeking, and that is all that it was, has put Durham University in a very bad light as well. We apologize for the charade."

"The many citings and supposed murder, all made up as well?"

Price hesitated, perhaps irritated by the unexpected questions, and then assured the reporters in a tired voice, "It is nothing more than acting, you know, performing for the camera. Some of the segments might have seemed real, but they were all contrived, designed to make a good story that would capture the imagination of the public, and it most certainly has."

"Who is responsible," another reporter asked. "Did Professor Parker pay filmmakers to put together this fake set of videos, or like you, was he fooled by the unscrupulous scientists?"

"It appears that Parker, from what I have been told, is a major player in the charade. We do not know for certain whether he funded the bogus tapes alone or with the help of others, but I have been informed by President Paul Knight that he no longer is employed by Durham University. This has been a terrible ordeal, an incredible embarrassment for all the innocent people who were duped by the publicity seekers. Please let me set the record straight. There is no Grendel, there never was a Grendel. There is no new evidence for a missing link or a parallel species. Now please let the matter rest. There is nothing more that I can tell you." Price, flashing his palms at the camera, waved off further questions and quickly left the podium.

Parker looked at Herder with an expression of complete shock. It was a brutal betrayal, beyond his wildest imagination, and like many other recent events in his life, entirely absurd. Several minutes passed before he could comprehend everything that he had just witnessed. It was beyond his most bizarre fears, his worst nightmare, for it truly was the unimaginable. Stunned, he placed his hand over his mouth and slowly shook his head. He could make no sense of the news broadcast. *Why hide the truth? So this was the modern version of what happened to Giordano Bruno. Nobody literally gets burned at the stake, with all of the screaming and charred flesh. No, that is much too primitive and certainly wouldn't go over well with a TV audience as it watched the evening news. Instead, one is simply permanently expunged from the dialogue. No voice, no credibility, no problem for the university or religious institutions. It was that simple.*

It was an ingenuous replay of George Orwell's famous novel where facts and documents were constantly altered, "updated" as Big Brother called it, in the name of "proper or correct understanding." And that is exactly what happened to Parker. Ten years of meticulously documented information suddenly became bogus because an authority figure said so. Of course Price presented no proof, just a simple assertion backed up by the weight of the university which the public assumed "had no apparent reason to lie." The individual, no matter how accurate their position, did not stand a chance. Who would believe them? There was no point to try to defend himself. Besides literally handing them a living Grendel, what could he produce that would provide satisfactory evidence? He had no recourse. He could get a lawyer, but that would only make him appear guiltier. *It seemed, once again, that the institution had won. Was there any way out?*

"I'm afraid it's just the beginning…things probably are going to get crazier, a lot crazier. I don't know what to tell you. I thought Knight was a decent guy…somebody must have gotten to him but that's something we probably will never know," Herder said as if thinking out loud. "If fanatics have one virtue, it's commitment to their cause, but I *never* thought in a million years that it would come to this, not at Durham University," he said, his eyes still glued to the TV, as if miraculously a rebuttal might be presented.

"I don't stand a chance, Steven. The academic world calls me a fraud and the religious world portrays me as a heretic. They make me out as against both God and truth. Who will believe me?"

"Now we see why it is so much easier to join them than to fight them. Price discovered that. I guess that is the great advantage of a fundamentalist religion or an academic institution with a preconceived agenda…no thought is required; everything is black and white with no shades of gray. Just be a good follower and the reward is yours. Nothing has to make any sense, just submit. What's worse, they make no effort to understand anything about your arguments, about how states of consciousness can manifest divinity or how innate truth, when viewed through the soul, can be sacred," Herder noted with a note of despair.

"It's such a paradox," Parker lamented. "The great religious figures, and all truly holy people for that matter, totally give up their egos and actually live in elevated states of being. They encounter the world with universal love. They judge almost nothing. Yet the institutions that have grown up around their name have empowered the ego, making each faith exclusive, dogmatic and unwilling to accept any other faith. For almost everyone, religion could not exist without the institution. Without a set of myths and rituals to glorify or a body of holy works passed down from generation to generation, billions of people would be forced to seek divinity on their own. For most people, I guess, it is much easier to conform to the damaged collective ego of a church, mosque or temple than to find their own way to God. And no one seems to want to recognize that religious institutions breed both hatred and bloodshed. There is a dark side to divinity," Parker said, looking off into the distance.

"Most of the world, Eric, is never going to get the idea of Christ-consciousness or Buddha-nature, the idea of actually living divinity. The institutions would lose all of their power if that happened and besides, it's too much to ask of the masses. Who has the time or the energy to actually try to live according to the Gospels when day-to-day concerns are so much more immediate?"

"The true experiencing of the sacred is not a power issue, Steven, or at least it shouldn't be. But when clerics admonish their followers to kill the infidel, when evangelists claim that 9/11 is the work of an angry God cleansing a corrupt city, when Israel practices an eye-for-an-eye, it is hopeless, doubly so when universities back away from the truth for fear of religious retaliation. It really is perplexing."

"Can you see a way out," Herder asked, his eyes searching for Parker's?

"If you look at the big picture, there is a glimmer of hope," Parker paused. He even smiled slightly. "A small, but growing number of people have found what I think is the only solution. Each of us must create our personal belief systems, ones that can be fully embraced and lived, ones that expand our reality rather than shut it down with massive layers of structure and judgment, ones that foster tolerance, compassion and love. We each have to directly experience the sacred in our own way. Steven, there are more aware people on this planet now than ever; they are the hope for a peaceful future because their souls can truly embrace the transcendent."

"Yet each religion claims that a personal creed, the kind that you just described, is the ultimate act of ego, of the individual asserting his or her will," Herder noted, "and it is a bit ironic, but in some ways it is true, is it not?"

"How can finding a personal path to a compassionate, loving Divinity be an act of the will, an act of the ego, when it can only be attained through egolessness, through totally letting go… then we are able to relinquish personal control of our lives and actually experience the transpersonal. When we are in that state of universal love, we give up our differences, our cultural identity for that matter, and embrace the sacred unity within each person and all of life."

"Wonderful idea, Eric, but the millions who follow that path will never have one hundredth of the power of the church or temple or mosque. The individual journey is largely anonymous and invisible. It requires so much more energy and commitment," he paused, looking out the window at the falling leaves, "even worse, you have to go it totally alone. Most of us are creatures of comfort, so we gravitate towards what already exists, particularly if millions of others are doing it and it offers guaranteed results, like heaven or paradise or God's approval."

As an after thought, Herder added, "And as far as love is concerned, that word extends only as far as the people in one's group: Christians love Christians, Jews love Jews and Moslems love Moslems. Each of them might claim something different, but generally it's a lie. In each case they are discouraged from marrying outsiders, or even having too close of a friendship. This type of prejudice isn't going to change anytime soon," he said in a wistful voice.

Herder glanced at his watch, "Sorry Eric," he said, picking up his briefcase and heading for the door, "I'm late for class." Alone again, Parker realized that his time at Herder's was running out. If he were discovered, Herder would be the next person to become persona non gratis.

Parker's attention was drawn outside, to the swirl of gold and orange blowing in the morning breeze. His life now was just as chaotic, his future just as uncertain. He intended to depart no later than midday, but that allowed enough time to view another sequence of the tape. He walked into the den, slipped the video into the tape deck and sat back. There was no need to take notes. His professional observations were of no interest to anyone.

"The Virtue of Stillness," a piece by Bronsky, first caught his attention. She asserted that most people live with continual noise which acted as a permanent barrier, separating each of us from the Ineffable. In fact, she claimed that the youth of America had never known utter silence and was dreadfully afraid of it. Contemplation, on the other hand, was the work of an active mind, one that was shooting words into the stillness, hoping to make meaning. Each word had an intention; it disturbed the silence in its attempt to denote truth. It was far better than noise, the meaningless drivel generated by the media, but it was still produced by the mental struggles within the self. The deepest state of grace occurred when the mind was totally empty, a void, with no intention. Yet how many people could truly understand that?

Parker agreed with Bronsky's assertion. Clearly it was at that moment that one penetrated through the veil of illusion, and with the help of ones intuitive right brain energy, grasped the true origin of things; here one entered a sacred silence where the presence of the Oneness could be glimpsed. It was a wordless union, what some Hindus called "the kiss of the divine," a perpetual processing of the world as if one embraced and was being embraced by God. Bronsky suggested that other than Grendel's infrequent interactions with mankind, he naturally enjoyed the splendors of silence and hence often encountered the presence of the Oneness. This resonated with Parker, who daily practiced meditation, and on occasion also blissfully experienced the wordless.

When his focus returned to the video, at first, the grainy black and white images were difficult to discern. Grendel was in semi darkness, perhaps resting in a cave somewhere. He appeared to be relaxed, drifting in and out of sleep. A sound close by suddenly grabbed his attention. He sat up; his head lifted a bit, as if he were trying to smell something. The monitors spiked, indicating that he was on full alert. Danger was nearby. Then appearing at the cave's small opening, the camera revealed six boots and three sets of legs from the knees down, no more than five yards away. Voices were calmly speaking a language that Parker had never heard. A pole was pushed several meters into the entrance of the cave, slid from side to side, then the party moved on.

After several minutes, when everything seemed quiet, Grendel unexpectedly exploded from the cave. *What was he doing?* Several shots rang out, the bullets wildly ricocheting off the boulders. Parker was stunned. Grendel was in full flight mode, ducking behind massive rock formations and sprinting at speeds in excess of thirty miles an hour in areas that were largely unprotected. An occasional shot broke the silence, but as Grendel made his escape, the shots became more infrequent and further in the distance. Eventually he stopped his mad scramble and rested in the safety of a small declivity. His heavy breathing was interrupted by a short grunt. The camera revealed a patch of redness on his right side. He was wounded, but thankfully, not seriously. He seemed to realize that the blood slowly dripping from his side would make him easier to track. With renewed energy he moved on, not in panic but with determined intention, as if he were aware that his survival was in jeopardy.

The next scene revealed Grendel at a lake, gently washing his wound, which appeared several inches long, but judging by the limited amount of blood, not too deep. He was cautious. His eyes scanned the nearby ridges and several times checked for activity along the path that led to the water. There was no movement. He followed the cleaning by applying large amounts of saliva to the opening. Satisfied, he quickly headed for the nearby underbrush and continued his journey down the mountain.

Flushed out from his customary habitat, Grendel was in the world that he often observed from his elevated perch high in the mountains, a world filled with man's stupid aggressiveness, a world that he successfully avoided until now. Parker realized that Grendel was more vulnerable than ever because there were fewer areas for him to take refuge and during daylight hours, a greater likelihood of contact with humans. It occurred to Parker that Grendel would be better off sleeping during the day and traveling at night, his black fur more difficult to detect in the darkness. Parker fervently hoped that Grendel would return to his nocturnal habits.

The tape showed Grendel scrambling through a pasture, the cows paying little attention to him, but he always kept his distance from the farm house. Initially he was successful in avoiding people. Perhaps his strong sense of smell was most useful in his evasion. The video was obviously edited, omitting long periods of travel through woods and farmland. Perhaps it was the next morning, but as he crossed a road, he startled a farmer who had stopped to adjust the vegetables on the back of his cart. There was what sounded like a curse, but Grendel moved so quickly that the farmer appeared to be unclear about what he saw. Was it a bear? As the day progressed, Grendel seemed more confident, as if he were certain that he lost his pursuers, and more curious, as if he were increasingly interested in the behavior of humans. Instead of staying secluded in the deeper parts of the woods, he spent more time in the fringe areas that divided the forest from the farmland. Here, he observed people from a safe distance. What he learned about human beings, if anything, remained a mystery. Ironic, Parker thought, *Grendel was spying on mankind just as Parker was spying on Grendel. But how much longer could he evade detection, Parker wondered?*

The sequences that followed seemed insignificant: farm hands working in the fields, cars and a few trucks driving on a gravel road, and an occasional scan of the tree line behind him to make certain that he was not observed. Grendel seemed comfortable and adjusted to his surroundings. As night fell and the temperature dropped below freezing, Grendel seemed agitated by the cold and confused about what to do. There were no caves to provide shelter, so Grendel decided to seek protection in a shed a considerable distance from a farm house. Apparently, he spent the night.

The footage began again in that grayish moment when darkness was beginning to subside. As Grendel was waking up, something was approaching rapidly. As he was leaving the shed, a large dark animal growled, barked and then attacked. Grendel rose up and in mid air, delivered a blow that left the dog unconscious, if not dead. Once again he fled, his heart pounding frantically. *Was he being tracked by his earlier pursuers?*

Not knowing the locale, Grendel could be more easily trapped. No massive boulders provided protection. The further he fled from his home in the high mountains, the more inevitable an unfortunate encounter with man. Parker had an uncomfortable feeling about Grendel's destiny. Although the videos were probably more than a decade old, Parker felt as if he needed more information. *Something seemed wrong!*

On a hunch, he rewound the tape to the gravel road sequence and paused on several trucks that seemed insignificant upon first viewing. One close up revealed part of a gun; it looked like a semi automatic weapon. Another enlargement exposed part of a military uniform and an emblem that looked like a clenched fist with a crucifix dangling from it, the type of symbol that he saw on the casket bearers at the earlier funeral for the fallen "hero." These were people from the high country, Father John's followers. *What were they doing here? Were they finally closing in on Grendel?*

# Chapter Eight: Flight

Before he viewed the final tape, he phoned the only person who might be able to help him. Fortunately Professor Bronsky was available and willing to talk to him. She was aware of the morning's events, both Price's condemnation of the project and Parker's sudden expulsion from the university. Because of her busy schedule, she had not accepted the invitations to appear on national TV. Although her future with the university was still undecided, she clearly sympathized with her cohort, believing like him in the legitimacy of the project. She made it clear that she was willing to help him in any way that she could.

Parker quickly came to the point. "I need to personally investigate, to find out what's really going on if I can. My future depends on it," he said hurriedly as if he already were running for his life. "Helena, the woman who contacted you regarding her adopted daughter, the one with the reoccurring nightmares, you must have some way of finding a point of origin, you know, where the journals were sent from."

"They came wrapped in neatly snipped paper bags that were held together with masking tape. Truthfully, I was curious. I can give you a country and city, but the woman never identified herself. Foca, Bosnia, is where they were mailed from. Do you know anything about that area?"

"Nothing."

"Back in the mid nineties, it was a blood bath...right in the middle of that ethnic cleansing mess. Half of the city was Muslim, the other half Christian. When it was all over, all the mosques, fourteen of them I believe, were destroyed and twenty thousand Muslims were either slaughtered or fled for their lives." She paused a moment, then added, "The brutality of the Serbs knew no boundaries. They turned the prisons into torture chambers and set up rape camps where young women were repeatedly assaulted. And Eric, all of this was done with the tacit approval of the church. Like so many slaughters, it was largely based on religion. You might remember Clinton was on the verge of being impeached over the White House aide scandal. To shift the limelight from his troubles, he sent the US Air Force to support the UN to try to quell the mass murders."

"So…Grendel was fleeing into one of the worst genocides of the twentieth century. Helena, you wouldn't happen to know what the outcome was, I mean, what really happened to Grendel?"

"If you mean, was he destroyed or did he survive, I can't help you."

"I need to find out… and I also need to get out of here. I'm going to fly over there, today, if possible."

Bronsky volunteered to contact her friend and former student, Stojan Janovic, who was educated in an American university and could act as both his guide and interpreter. Parker was in no position to decline. Before the conversation ended, he graciously accepted all of her assistance.

As he was reaching for the last tape, the phone rang. *Perhaps it was Bronsky?* He picked up the receiver and was confronted with a voice that he had heard before, the same distinguishing Middle Eastern accent of his assailants. "We know where you are. You are going to die today." Then a loud click that might as well have been a gunshot. Parker panicked! Thoughtlessly, much like Grendel in the tape when his survival mechanisms were triggered, he allowed his primal instincts to take over. He immediately called a taxi, gave the location of a nearby McDonald's, grabbed the tapes and his suitcase, and was out of there. He hurriedly walked the four blocks, constantly looking over his shoulder, expecting the worst. He didn't want to call attention to himself, but he was terrified. His heart beat wildly, thrusting his body in full flight mode. *Was the gorilla masked man nearby?*

He tried to remain calm while blending in with the other mid morning McDonald's regulars. For the first time in his life, he experienced a total alienation from the culture that had nourished him. Because he was displaced, he saw the world differently. Parker ordered coffee, although he detested it, and sat at a booth that allowed him to casually observe the entrance and parking lot. No taxi yet. A smiling Ronald McDonald caught his eye. Like everything else around him, the clown seemed trite and sterile. *Why does America glorify plastic? There was some truth in the mullah's condemnation of our culture,* he thought.

He gazed across the street. The dozen stores at that small mall could be found in almost any other city in America. It occurred to him, *no doubt we learn to buy our identity and then, year by year, upgrade it as our image requires and our income permits. The mullah insisted that our real religion was greed: the consumption of what we don't need to create a false picture, a fantasy of sorts. Was he correct? Was America more illusion than substance?* His eyes searched for something real, but everything was flimsy, from the artificial plants to the plastic seats, all disturbingly fake. Even the food, whether a Big Mac, super-sized fries or a giant Coke, seemed unnatural, almost as if they were fatty pellets designed for rat consumption. Dehumanized, he felt as if he were in a giant cage that was not even suitable for rodents.

For years he was so busy with his career that he neglected to notice the obvious. America was for sale! Truth was irrelevant. For most of the US, an enjoyable illusion seemed preferable to a meaningful, but less spectacular, reality. For a moment he felt disoriented, even to the point of doubting his long held convictions. *Could Price be right? Or were the trusted voices of authority just tools, perhaps unwittingly, for the agendas of the institutions? Could it be that Grendel was a myth, just another hyped up lie? Is there anything left that is believable?*

Still no taxi.

His mind most of the time seemed to be one immense question mark. *Can one construct an authentic life in a culture that has no substance? Clearly, only a genuine existence can embrace divinity. How can we immerse ourselves in a plastic world, spend ten hours a day on computers in skyscrapers, fill our brains with hundreds of glossy images, each a clever misrepresentation, and still believe that our lives have meaning?*

Parker wanted to defend his culture, but at that moment the absurdity of America's lifestyle could not be justified. Certainly, he thought, we do have religion in the traditional sense, but the sacred, the major purpose of a faith, seemed to have dried up like summer flowers in late August. What remained were millions of people, spending one hour a week in a place of worship, attending to the holy almost as if it were a job or some social engagement. In the end, it seemed just another duty to be performed. He remembered reading a quotation that suggested that formal religion was like going to kindergarten, while true maturation produced a deepening spirituality.

At last, the taxi arrived.

He scurried out to meet it. The driver, a Muslim wearing a colorful turban and speaking broken English, shocked Parker. Should he have expected something different? Once his suitcase and satchel were secure, Parker spoke in a controlled voice, successfully hiding his fear, "Airport, please."

The trip was slow, the driver choosing to listen to a foreign music radio station rather than converse with him. He tried to repress his fears, but every so often his eyes would meet the taxi driver's eyes in the rear view mirror. *Would there be a sudden turn into a back alley? Was he going to be delivered to the gorilla masked man?* It was all that he could do to remain calm.

The airport, three large terminals connected by people conveyers and a monorail, was a jumble of activity. Thousands of faceless travelers seemed to be moving in countless different directions. He was surrounded by the impersonal feel of glass and metal, a modern day cathedral of sorts preoccupied with images and illusions. Dozens of large, colorful advertisements, with beautiful, skimpy-dressed women and picturesque vacation destinations, created a sterile, fantasy world that was devoutly worshipped by the majority. *Our gods, Parker thought, what we actually pay homage to, were money, beauty and the allure of escape.*

As he walked by a large, electronic board displaying departure times, he felt happy to be leaving. For the moment, like almost every expressionless face that he saw, he felt as if he didn't really exist. He was nothing more than an anonymous figure in an indifferent universe, a miraculous substance that could take on any shape required. His thoughts turned to the Jewish philosopher Martin Buber, who characterized modern man as nothing more than an "it," a thing, empty of meaning and with no true center of being. Parker agreed with Buber; most of us had become depersonalized objects to be used and then discarded. We all had to perform, become the job description or the appropriate image, or our future was in jeopardy. There was no place for our humanity.

He listened as the upbeat background music serenaded the crowds of empty people, largely soulless wanderers, he thought, floating on conveyor belts to nowhere. In this hyperactive pleasure dome of sorts, any genuine thoughts of the sacred were as ridiculous as the idea that in the next waiting area Christ could be heard repeating the Beatitudes. There was absolutely nothing holy about America's daily way of life. As the all pervasive audio system barked directions in a numerous different languages, Parker thought *this really is the tower of Babel, and we are a multitude of antlike "its." Trapped in our false identities, we are quietly moving towards oblivion, each of us, as Pink Floyd noted, "comfortably numb."*

Shortly after six, he boarded his plane to Frankfurt, Germany. Because he could never sleep on transatlantic flights, he expected a long night. For the better part of the first hour, he enjoyed watching the changing topography from his window seat. Soon, the sun set and darkness quickly covered everything, including his thoughts. The movie, an idiotic comedy about an incompetent family that lost their pet snake, was no help. As he laid back his head and shut his eyes, he remembered frustrating discussions with well meaning students. Harsh voices of dozens of young men and women reverberated in his tired brain, a barrage of angry remarks, hasty rebuttals that challenged everything that seemed so obvious to him.

"Transcendence has nothing to do with religion, Doctor Parker. Why would a change in consciousness, moving from one level of reality to another as you put it, have any connection with God or religion? Why are brain waves so important to you? God is in church, in the Bible, but brain waves, you've got to be kidding me."

"Yes," he heard himself saying, irritated that they missed the obvious, "but how many people go to a place of worship or read the Holy Scriptures, claiming to be religious, yet no transformation occurs within their being? They dutifully carry out their performance with no internal corresponding change in consciousness. Religion **must** be a direct experience, not just a set of beliefs or a learned point of view. In our culture, ordinary reality is far from sacred. If a person doesn't get out of this work-responsibilities-social expectations state of mind, and therefore change his or her grasp of reality, then that person will never genuinely experience the higher states of being and what it **really** means to be a spiritual person."

"Nobody can live in a state of transcendence, and if they did, they would be locked up and put in an institution for the mentally unfit," another voice bellowed, as if calling from the top of a corporate headquarters, "We can't afford to have people waste away their lives…they need to be productive, to do something meaningful and to contribute to our society. Now tell me, how does transforming brain waves make a better America and improve our quality of life?"

"That's the point; don't you see that it makes all the difference in the world. How you experience what you are doing is **everything**! The brain waves don't lie; they are the real indication of your experience. A deeply depressed person functioning in a dark, suffocating frame of mind is completely alienated from the highest states of being. Isn't that obvious? Their caged ego can think about God and even perform holy rituals, but that is no guarantee that their soul will ever know the warm light of the truly sacred. Conversely, alpha waves create a very different way of embracing life. Because they are not a product of the left brain, they peacefully open the doors of understanding that lead to the divine. Can't you see that a change in consciousness puts you in touch with the highest, what is really there and has been there all along?" Parker's words became more strident as he tried to convey the obvious.

"Think about it. If you are suffering and deeply depressed, even feeling indifferent and thoroughly bored, if asked "how are you doing," you might say you are "fine." At that moment, however, you are in total denial. Is it not the same with religion? You can talk God or salvation from a left brain reference point, but that is far different from being unconditional love or being truly compassionate. Brain waves express your state of being. Your consciousness, not your thoughts, is truly who you are. That is your true reality. Divinity is not conceptual and centered on ideas: it's all about the level of awareness with which you process your experiences."

"Transcendence is not normal," a woman's voice whispered. "Reality is all those ordinary things that we do every day, mostly because we have to. A religious person is one who is true to their duties, making good moral choices. It is the act of willing oneself through your mind to carry out God's plan as stated in the Bible. States of being are irrelevant. A righteous life is everything."

"It's impossible," he responded. "Most people require literal direction. But what if Christ did only his duty as a good Jew or Buddha only lived as a good Hindu? It is such a paradox. They both moved beyond the older tradition, beyond what they saw as flawed dogma, yet they are revered as avatars. Ironically, when someone else works beyond the dogma, they are a threat, or even heretical. Some people get it, though. At least five million Americans meditate on a regular basis, many of them experiencing at least some change in consciousness."

"So what," a voice countered. "Prayer is how you really speak to God."

"Prayer is ego driven," Parker answered. "It's suggesting that the way things are is a mistake...God goofed up, and that our sense of how things should be, our list of concerns, that's what God should implement. But that way, nobody would ever die, get a life threatening disease or suffer for that matter. Prayer can be viewed as just another example of the mind seeking control, unable to let go and live in harmony with what is. Meditation, conversely, is the elimination of thought, an empty mind simply merging in the present with the Universal Energy." For him, it was self evident, although not so simple to carry out. Frustrated, he wondered *what good is teaching if there is no transformation?*

As the hours passed and the flight droned on, the angry debate gradually subsided. His thoughts turned to natural man and why all the traditions of Abraham vehemently condemned him. Obviously if natural man were desirable, there would be no point to formal religion. Helping man regain his birthright of childlike innocence would render the institutions pointless. To suggest that there is a beautiful, God-generated world beyond our ego is unthinkable. Although some religions have glorified to some degree the Edenic state prior to the Fall, their purpose was not to seek a return to primordial perfection. Rather they emphasized that having lost that state, man is primarily wicked and must seek redemption. For them, the road to heaven was an external process that demanded conformity to revealed dogma and myths.

Parker resisted this indoctrination. In fact, he saw Christ as the epitome of natural man, a heretic who rejected most of his culture's tainted teachings in favor of two natural principles innate within man: love of God and love of fellow man. Our deepest drives, Parker believed, were the need for an intense spiritual bonding with the Source as well as with the Source as it is manifest through humanity on the whole. Only our egos separated us from both.

Over the centuries Christianity had portrayed natural man as an "animal" who was preoccupied with the flesh, not the spirit. The Bible claimed that he was "an enemy to God," because he "resisted God," and ultimately was "without God." Yet, he needed God's enlightenment. Parker knew all the arguments. However, the truth seemed self evident. Man was born with a pure soul that was corrupted by the society in which he dwelled. Man's task was to return to his original state of perfect oneness with the Creator. The Taoist called it "letting go" and "going with the Flow," while the Buddhist claimed it was man's inborn capacity to actualize the Ultimate Truth, thus living in nirvana. Parker deeply desired a return to this state of innocence, but at the same time recognized the multitude of obstacles. Historically, very few people have had the self discipline and courage to go back to what the Taoists called "the uncarved block."

Eighteen hundred years ago Christianity violently obliterated the Gnostics, those of their own sect who believed that everything, including God, truth and morality, was found within man. The church brutally tortured and killed thousands who devoutly worshipped Christ, but not in the "correct" way. The Gnostics knew that God was beyond the ego. Adam, the father of mankind, was pure and wholesome until he ate of the fruit from the tree of knowledge of right and wrong. He disobeyed revealed religion and did what Parker was doing, seeking truth on his own. Adam's mistake, perhaps, was that his seeking was prompted by his ego, not his soul. Parker realized that most of the wisdom in *The Lustres* was heretical, validating a pre Fall Adam or the intuition of Christ or the enlightenment of The Buddha, all sources of natural man. Clearly, Christ's relationship with his Father was not based on a prescribed religion, but generated from an innate, personal understanding.

He knew that formal traditions could never embrace a regressive process that claimed that man innately had access to the pure energy. If all we needed to do were go back to what has always been within us, formal religion would lose its power. In returning to the Source, we would have to remove the many artificial and often very corrupt layers of civilization. Churches, temples and mosques, as well as governments and systems of education, instead of being looked upon as man's crowning achievements, might be seen as malignant growths in an otherwise perfect body. *Could we ever replace the instructions with what the Taoist call is-ness, the natural flow of the universe?*

Paradoxically, a creature like Grendel, who clearly lived the Buddhist maxim, "first, do no harm," would then become an avatar, a model for mankind to emulate, almost Christ like in his simple innocence. With no socially constructed ego to fill his brain with hatred and a thirst for power, Grendel would function as his Creator intended. *Could pure energy, following the spontaneous prompting of our deepest self, be the key to Eden, a stairway to paradise? Could it be that simple? Can we only enter God's kingdom as innocent children, or as Grendel?*

Parker thought about the future of his two sons. He believed that all children before "the Fall," before being forced to play society's corrupt games, were inherently Christlike. They were pure spirits living in the here and now. Often he had witnessed that quality in his boys. They were not duplicitous, dishonest or manipulative, but genuine and trusting. *What was the world going to do to them,* he wondered? And over the years he had watched Kathy become a conventional housewife, a working mother who was driven by society's view of perfection. She sought much of the "stuff" that the media had to offer. Over time she had imposed on their sons many of the messages that her culture endorsed, the very burdens from which Parker was attempting to escape.

As one would expect, she wanted her sons to be model young men and her husband to play by the rules. Parker knew that society demanded performance, not authenticity. He could not save his sons, or for that matter, very much of himself. Everyone inevitably had to confront the darkness on their own. He was running for his life, ironically, from God-loving people who wanted to hunt him down and kill him like they might a rabid dog, all because, like Adam, he insisted on thinking and living for himself. He resisted putting his wife into that category, the group who wanted him dead, but by forcing him to act according to society's expectations, she was imposing another form of death on him.

*Before most children graduated from elementary school,* Parker thought, *the majority of their innate goodness had already been replaced by a callous shell that allowed them to endure the dehumanizing process of becoming an adult.* Already Matthew, his oldest son, had begun shutting down. His responses to the world had become guarded and at times automatic, as if the person's reaction to him was more important than the truth. Mark was sure to follow.

Of course the soul could never be destroyed; but it could be, and often was, simply disregarded. Perhaps that was the way that it always had been, not just in America, but in every society: detachment from the soul at an early age. Males usually were the first to succumb. Unfortunately, spiritual death was inevitable. Ironically, the bleak journey into the night, what one might call submitting to the repressive demands of adulthood, was an essential part of the soul's development.

Miraculously, for some special people out of the darkness emerged a magnificent rebirth, the ultimate resurrection, **a spiritual transformation that empowered true living**. Often it was in the form of a religion or philosophy that actually expanded one's awareness and offered through the **direct experience** of the sacred at least partial liberation from the numbness of a crumbling culture. Yet most Americans were too busy making a living to think about this type of self actualization.

For Parker, it was natural man's right brain qualities, traits like unconditional love, creativity and genuineness, which offered a meaningful way out. As he pointed out in *The Lustres*, true freedom could not be bought or acquired from without. If one were to escape the fetters of a left brain and a largely soulless culture, liberation would be solely a product of one's internal commitment to personal transformation. Yet, there were no guarantees. That's what made the Grendel Project so interesting. Grendel had no religious institutions to counteract; thus, his right brain dominant being naturally encountered divinity simply because he had an open channel to the Source. No doubt it could be that way for homo sapiens as well.

It was still early morning as the plane approached the Frankfurt International Airport. The red roofs of the houses in nearby small hamlets shone in the sunlight. Although Parker was able to sleep only sporadically, the sunrise kindled a feeling of renewed energy. He had escaped! For a few minutes he experienced an exhilarating freedom that grew out of a sense of abandonment. Everyone around him had purpose and direction. Only he was riding a wave of immense uncertainty, and that insecurity made each moment more vibrant. He felt truly alive! He had released, if only temporarily, much of the emotional and psychological baggage that he had accumulated over a lifetime. For the time being, he could be exactly who he was.

After several hours Parker was in the air again, occupying a window seat on a Boeing 737. A few hours later he landed at the Belgrade Nikola Tesla Airport, and with little difficulty, picked up both his suitcase and satchel and headed for the car rental area. He selected a 2006 Opel from the Red Line Auto Agency, had a quick snack and then was on his way. He traveled west out of Belgrade, a city of one and a half million people, encountering only minor traffic problems. There were a few areas of congestion and one major construction slowdown. By noon he was in the countryside, slowly making his way toward Foca where he hoped to find the final clarity, the indisputable truth.

As he drove through small hamlets, there was a bleak grayness to the villages; the homes were wall after old wall of patched up cracks that seemed, like people's lives, to be just holding together. Everywhere he looked, people seemed to be beaten down, almost as if they were sleep walking. They were enduring in a state of semi-darkness that suggested that they were losing the battle for life. This certainly was not the glorified poverty that he associated with a devout life. Clearly it was not the voluntary asceticism that the holy men had written about. Life here offered no liberation. There was no hint of transcendence, only hard faces, usually deeply wrinkled, withstanding the daily challenge of a harsh survival.

Often the church was the largest building in these villages. But just as often, it seemed as old and outdated as the culture that it symbolized. He wanted to avoid judgment, but the dazed, stoical appearance of those he saw proved to be too discomforting. He felt everywhere a muffled anger or what Thoreau called "a quiet desperation," a helplessness that came from the sense that life had been denied. Unlike in America where people attempted to hide their depression, here the masses wore their anguish on their faces, as if everyone were suffering from a permanent toothache or lifelong flu.

He approached Foca, now a city of twenty-five thousand Christians, with the hope of resolution; he didn't know how he was going to find the truth, but he had to try. At first glance the town of red roofs and stucco buildings appeared serene, belying its recent past. As he slowly drove around, Parker wondered *where should I begin? Could I locate the small mountain town viewed in the earlier tapes, or perhaps discover Father John's school and have a chance to interview him? Perhaps I could locate KA, the young painter, by finding his drawings at a market or maybe I could discover the church where the anonymous dream girl was dropped off? A large town couldn't bury all of its secrets. But one thing was for sure, it was not only pointless but probably dangerous to scour the distant mountains for Grendel.* Although there was no really promising lead and no clear place to start, he had to do something for his own personal satisfaction. Besides, the world had a right to know, even if the news, like Darwin's theory of natural selection or Einstein's theory or relativity, was uncomfortable, if not revolutionary.

# Chapter Nine: Slaughter

After a ten minute search, driving on narrow, bumpy cobblestone streets that never were quite straight, Parker found a small hotel two blocks away from the busy marketplace. Everything was packed too close together. He felt claustrophobic. After squeezing the car into a narrow parking space, with suitcase in hand and his satchel with the Grendel tapes over his shoulder, he walked into a five hundred year old building that by American standards would be described as shabby at best. He felt as though he just landed on another planet; clearly, he didn't belong. The clerk behind the desk as well as the solitary man seated in the small, dingy lobby viewed him with suspicion. What was he doing here? Was he another journalist investigating the atrocities that occurred here almost a decade earlier? Was he going to add to Foca's deplorable reputation in the court of world opinion? After five minutes of awkward communication-the clerk knew a few words of English-he had a room number and an old key. There were no elevators. He climbed to the third floor and started down a dimly lit, fairly narrow hallway. At least the room numbers were large and legible.

As he stopped in front of his door, fiddling with the key in the uncooperative outdated lock, a man exited a nearby room and walked towards the stairway. He smiled, his teeth appearing unusually white beneath his black moustache. As Parker moved to give him room to pass, suddenly the man lashed out and severed his shoulder strap with what appeared to be a box cutter. Parker dropped his suitcase and with two hands wrenched the satchel away from the attacker who then grabbed his suitcase and fled. Parker yelled and ran down the hallway, but the man exited onto an emergency stairway and into a dark alleyway. Parker realized the futility of pursuit.

If he caught up to him, what then? Returning, he opened the door to his room and sat on the bed, stunned. He felt a dull pain on his right side, just above his waist, and noticed that he was bleeding. Somehow he had been cut, but fortunately it was superficial. There was no bathroom, no medical supplies, just a dirty, discolored sink. First he gently washed the wound, and then he pressed a towel on it, stopping the bleeding after a few minutes. Luckily, it would not require stitches. Certainly, he could not report this to the authorities. He couldn't speak the language and even if he could, why would they help? It was obvious that he was an intruder. Registering as an American didn't help either.

Then the irony struck him. Because he held onto the truth, not letting go of the satchel with the video tapes, he shared the same wound as Grendel. Now they were brothers of sorts, both innocents wrongly treated by mankind. *What did they do to deserve this fate?* He realized that he could die in this small room on the other side of the world and nobody would know. Although his actions were not totally altruistic, he chose to believe that his suffering was a worthy sacrifice for the good of humanity. He thought of Prometheus who was chained to a rock by the gods and daily endured the opening up of his side and the removal of his liver, just to have it regenerate over night. The same painful process repeated itself over and over. Parker, on some level, thought that he too was stealing fire from the gods so that mankind might be able to have the darkness illuminated. Grendel might be the last of his species; even more significant, he might offer mankind a new vision of what the sons and daughters of Adam might have become: a powerful paradigm that glorified a right brain, largely intuitive-based existence, one that provided a direct path to divinity. Exhausted, he rolled back onto the bed, closed his eyes and hoped that rest would bring a new perspective to his evermore hopeless plight.

Before he drifted off to sleep, he heard the mullah's voice say, "There is your natural man, Doctor Parker," obviously referring to his assailant, "He is a man of the flesh who commits senseless evil. What kind of fool can hold him up as divine?" He heard himself answering back in a firm voice, "He is unnatural man, corrupted by the sickness of his society. What is natural about committing violence on a total stranger? That is not a behavior that man is born with. That is not an expression of his soul. It is, however, the product of the numbness that engulfs man when he is consumed by the distorted messages of his civilized, ego-driven world." Gradually, sleep prevailed.

The early day sunlight burst through the white curtains, casting golden rays of hope on the worn out walls of his hotel room. He was just concluding his morning meditation, a deep, hour long session that brought him into a space of peace and clarity, when a soft knock on his door interrupted him. Very cautiously, he opened the door several inches and to his surprise, looked into the blue eyes of an American-dressed man, perhaps in his late twenties.

"Hi, I am Stojan Janovic. Doctor Bronsky suggested that I could assist you." And then in perfect English, "My friends call me Jan." Parker, happy almost beyond words, opened wide the door and introduced himself. Then he asked, "How did you find me?"

"There are only three hotels and this was the only one that had a rental car parked in front of it. I told the clerk that I was to meet an American friend. He gave me your room number," Jan said with a smile that suggested that he was pleased with himself.

Then Parker described yesterday's misfortune. During the evening, after some deliberation, he decided to let the matter drop. The tapes, not his clothing, were by far the most important items. Going to the authorities would take too much time from his investigation; there was no point in formally pressing charges. Parker was excited to learn that Bronsky had provided Jan with two possible areas of interest, both nearby and both extremely unusual. After a filling breakfast of dark bread, cheese and several types of cold meats, they were off to the only museum in Foca which sat on top of a small hill.

The large, stone building, three floors in all, was about the size of a soccer field. It was divided into four different compartments. Jan hurried him towards the Natural History section which required them to walk down a long flight of poorly lit steps. They took several wrong turns as they followed dark hallways leading nowhere, unnaturally blocked off by recently constructed cinder block walls, and generating the unsettling feeling of a Kafkaesque labyrinth.

Eventually Jan found the office of the director, a slight man in his mid fifties, who was dressed in a musty, white lab jacket. After a brief debate and not without some difficulty, Jan managed to get access to a room that was securely padlocked. A half century ago, it was part of the museum's standard collection; now it was clearly off limits. With the advent of science fiction and horror movies, the old exhibit proved too disturbing for many of Foca's younger patrons. The director explained to Jan that visitors, mostly children from the nearby mountain villages, were especially affected by its contents.

At first, Parker saw nothing unusual, rows of what appeared to be human bones neatly separated and ordered on four long tables. On the walls were three rows of shelves containing at least forty human skulls, each perfect, as if carefully manufactured in a nearby factory. Parker thought that these people had to be vegetarians and innately peaceful to have such little evidence of violence. Often skulls from other parts of the world displayed long cracks or gaping holes that suggested an unpleasant death. At the far end of the room was a skeleton that stretched from the floor to the very top of the eight foot ceiling. Its head was bent forward and its unusually long arms protruded out into the room as if it were a mannequin from a sleep walking scene in an old Frankenstein movie.

Then it occurred to Parker that he was looking at what might be a close relative of Grendel. Everything in the room was super-sized, from the ten inch fingers to the hands as large as a baseball glove, all massive, more than double the size of an average man. The strength of the creature was beyond comprehension. Parker felt unsettled by it all, as if witnessing the ghost of a real monster. In truth, the exhibit could easily terrify children, especially if they heard embellished stories about a living Giant from the Darkness, one lurking at night near their house or in the shadows of their imagination. Parker began to inquire, "How did the museum acquire such an unusual collection," but the director was emphatic. "No, enough," he shouted at Jan. They had to leave, immediately!

Jan explained how people in the area were sensitive about their history. Filmmakers had come through looking for Noah's ark and other artifacts mentioned in the Bible. Many of the peasants in the small villages were uncomfortable with the intrusions. They did not want their heritage looted and bandied about. The director claimed that what Jan and Parker briefly saw was evidence of the race of giants that lived in the area thousands of years ago and were frequently referred to in the Old Testament. The last thing the people of Foca wanted, according to the director, was to bring further notoriety to itself through another bizarre set of events. They had enough bad press.

As Parker and Jan retraced their steps back to the car, Parker recalled how in the Bible the giants were never spoken of as "God's people," but always existed on the fringe of mankind and often associated with some implied darkness, as if subhuman and perhaps soulless. The fates of Sampson and Goliath were far from heroic, neither story eliciting much compassion from the average Bible reader. Interestingly enough, the holy book suggested that being too big, having too much flesh, was somehow evil, whereas a big heart or a well developed brain was not condemned, but often glorified. In the Old Testament it was clear that giants existed before Adam and Eve, but obviously God was not satisfied with them; hence, a new race, mankind, was formed, "made in the image and likeness of a more diminutive God."

Furthermore, God's chosen people, the obvious writers of the Old Testament, now were the only valid source of His word. Parker wondered, *were the giants despised and destroyed because they were remnants of natural man? One thing was clear, however: when natural man came in conflict with adaptive man, who now occupied well over ninety nine percent of our planet, natural man lost the battle every time. Because of this, he was relegated to desert areas and mountainous regions of little interest to industrialized man. Soon all that will exist of primordial man will be photos from early editions of National Geographic magazines and some old film footage tucked away in cobwebbed archives. His ways of knowing, his "medicine or magic" as it was called, all quite different from modern man, will be gone forever.*

As they walked into the parking area, Jan explained how some people thought that these creatures still existed in the mountains east of the city. The peasants had mixed feelings about them. Although there was no evidence that they attacked man, some farmers hated them. Apparently they were terrified by an animal of such massive proportions. Others tolerated the infrequent encounters and reacted to them almost as if they were a nuisance, nothing more than oversized bears. Jan concluded, "Very little is known about them, if they exist at all. The mountain people are very secretive to begin with. It is unlikely that we are going to learn much more."

Next Jan led Parker to the city prison, a large building with iron bars over all of its windows. Jan informed him that the dungeon-like edifice was used as a torture chamber and "death camp" during the ethnic cleansing of the nineties. Part of the facility was still used as a jail, but the other half was now a mental institution. A stern faced, overweight woman greeted Jan in a matter of fact tone of voice, asking him to state his business. He requested to speak with Doctor Vladimir Dienje, head of the psychosis ward. After a short wait, they were led to a small room that looked like a principal's office in a 1940's rural school. Jan explained to Dienje that Doctor Helena Bronsky had read about one of his patients, case 162, and that they were interested in learning more about him.

Dienje, who had published several books and apparently knew of Bronsky, explained that the young man, now in his late teens, was unable to satisfactorily function in normal reality. His psychosis, a constant interaction with a non existent person, had not responded to any form of treatment. He still spent much of the time talking to an imaginary friend, often disregarding the world around him. Unfortunately he had been in various institutions, mostly orphanages, since he was found walking the streets of Foca about ten years ago. Although nine or ten at the time, he could provide no identification and when asked, could not remember his own name.

He repeated the same story continually, that he was saved from certain death by a giant, superhuman creature with three eyes that he affectionately called Dazbog. The story never varied. The frightened boy was hiding. There was total mayhem all around him, gunshots, blood, screaming, and just complete pandemonium. In the middle of all this, a giant arm unexpectedly grabbed his body and lifted him to safety, "floating him far away." Dienje noted that "All day long he talks to this imaginary Dazbog, much as a religious person would talk to God. Sometimes he asks Dazbog questions, but mostly he thanks him over and over again, repeating, 'I am grateful most powerful lord.' To put it bluntly, he is totally obsessed with this imaginary creature which seems to inhibit any type of normal functioning."

Parker asked Jan to request a visit with the young man. After a moment of hesitation, Deinje agreed. Parker was happy that case 162, now called Alex, was to be delivered to the small room. He didn't want to see the horror that he imagined existed behind the padlocked door. The young man, large and dark complexed, arrived quickly and in good spirits. Although shabbily dressed, he was remarkably well composed. After Alex was seated and comfortable, Jan translated Parker's questions and the boy's answers.

"We are interested in your Dazbog," Parker said. "It might make you happy to learn that we know some other people who have met him as well. Now this creature who you talk about and talk to, it wasn't a human being, but something that in some ways looked like one. Is that correct?

The boy listened as Jan translated Parker's words and then spoke a single word.

"Yes," Jan translated

"And he saved you from a horrible death?" Parker asked.

"Yes," the boy said again..

"Did he kill those who were going to hurt you?"

"No," Alex said calmly.

"Just helped you escape."

"Yes."

"Did he talk to you?"

"No," he said, shaking his head slowly.

"Did he save any others besides you?"

"I'm not sure," He paused a few moments. "Maybe."

"Did you know those other children?" Parker asked as if it were a fact. He scratched his head. "Maybe."

"Were you afraid of him?"

"At first, but he was very kind," Alex said with a slight smile.

"Did you trust him?"

"Very much!"

"More than people?"

"Yes!"

"Why?"

"Because he protected me…I felt safe."

"Does he still protect you…even after all of these years?"

"Yes, he still watches over me. I feel him right now."

"I'm going to show you a picture drawn by a young man who feels very much like you. Tell me, is this what Dazbog looks like?" Parker then showed him KA's drawing.

After a long silence Alex wiped away a few tears. "That's him, exactly."

"What do you see in the middle of his forehead?" Parker asked.

"A magic eye that sparkles in the sunlight. It sees everything the way things really are."

"Why do you still talk to him?"

"He is the only one who tells the truth, the only one who I can trust," he said looking squarely into Parker's eyes.

After a moment of hesitation, Parker asked, "Do you believe in a God?"

"I believe in Dazbog."

"Because Dazbog is the only one than can make your life better…the only one who can save you from suffering?"

"I guess," he said after a slight hesitation.

Parker realized that the questions were getting too complex. "Thank you Alex for your help. Is there anything that you would like to ask?

"Could I have that picture of Dazbog?"

"Gladly, I will make a copy for you and give it to Doctor Deinje."

When they arrived at the car, Jan asked Parker, "What's going on?"

"I'm not sure, but Dazbog seems to function as a God for Alex, offering both protection and hope. The answer, if we are lucky, might be on the next tape. Do you know any place we could go where we could view a videotape, not a CD?"

"The library might have a viewing room."

As they drove to the other side of Foca, pass dozens of old dwellings, many with an exterior of cracked stucco, Parker imagined the people existing on the other side of the dark, old walls. Seated around a kitchen table or lounging on a couch watching a small TV, were people who thought of themselves as good Christians, people who regularly attended Sunday Mass, people who knew the Bible and sometimes read it. And yet, paradoxically, most of these men and women embraced ethnic cleansing, the cold-blooded slaughter of hundreds, if not thousands, of people who were living similar lives, but had a different belief system. As it turned out, Parker realized, it's all in the Bible and the Christians of Foca, in some respects, performed much better than God's chosen people. Jan wasn't particularly religious. He knew almost nothing of the Old Testament. Parker slowly related the Biblical story that troubled the professor more than any other.

"Jan, you probably know that in the Bible God manifested traits that He strongly warned people to avoid. God, at times, is described as angry, even vengeful, and in other verses as envious. Even worse, He instructed his 'favored people,' which seemed to me contradictory because God created all life on this planet, to kill 'every man, woman and child' in a neighboring tribe, but the chosen people disobeyed Him, killing only the men and mercifully taking the women and children as slaves. You might wonder what this nearby tribe was doing that so offended God. As it turned out, they were referring to

Him by a different name and worshipping Him in an inappropriate manner, as pagans, seeing God's reflection in nature. In any event, God was deeply angered that his chosen people did not carry out His instructions."

"That's really in the Bible," Jan asked? "The God who instructed the chosen people to kill in cold blood is the same God who gave the commandment, 'Thou shall not kill.'"

"It's confusing, isn't it? Killing was OK if God instructed man to do it, but not OK for man to decide on his own. So what is really right? One could claim that the good Christians of Foca were following a three thousand year old Biblical story that justified the slaughter of people who worshiped God differently. They decided that the neighboring tribe, Muslims, were offending God in their inappropriate rituals. So as usual, we have good, God-fearing people carrying out what they think is God's will as it's stated in the Scriptures." Parker paused, as a new idea occurred to him.

"Funny though, if you take institutionalized religion out of the equation, the result probably would be quite different. If war broke out, it would be over a much more concrete matter...not some 'holy' words. And I doubt whether there would be genocide, but if there would be, it would not be carried out in God's name. In fact, ethnic cleansing might be nothing more than," after a long pause, "perverted religious socialization. That's what's so perplexing to me."

The library was able to accommodate them, providing a cramped room with an older model TV. Parker, anxious for answers, quickly reviewed the most important recent events for his startled guide who knew nothing about Grendel because the European news outlets did not find the Grendel Project news worthy. The tape began with Grendel completely hidden in the underbrush, viewing the world from ground level. A small band of men, about a dozen in total, moved quietly through the meadow, their thick boot prints leaving heavy indentations on the field of grass and blotting out golden clumps of mountain flowers. Most were armed with rifles; one or two had semi-automatic weapons. Parker immediately recognized them, not just because of their insignia, the clenched fist with a cross dangling from it, but from their semi-military outfits as well.

These were men of Father John's congregation, many of them familiar faces who were videotaped at the funeral. *Were they finally closing in on Grendel?* Grendel, with his vital signs almost off the charts, remained completely still, nothing more than a dark log covered with foliage. On the other side of a nearby stream, the group silently divided, the leader using hand signals to point the way. *What military operation were they engaging in?* After five minutes, the pathway totally quiet, Grendel, who had become more and more inquisitive about the activities of human beings, cautiously followed.

In the distance there were sounds, at first barely audible, as if people were singing. As Grendel got closer, Parker realized that it wasn't music, but high pitched shrieks, like when one of the stalkers fell off the cliff. Grendel continued in the direction of the noise which now included occasional gunshots. When he reached a clearing, five or six houses and all of the smaller buildings, the whole hamlet in fact, was burning. As he viewed the flames, he suddenly noticed several men lying motionless in the road, their bloody heads almost completely severed from their bodies. Someone had slit their throats like farmers butchered pigs and cattle. Several men stood on the outskirts of the village, their rifles in a ready position, shooting anything that moved. Grendel watched as they methodically gunned down a man in his early twenties, a toddler and three dogs.

Perhaps he remembered the drunken shooters who were torturing the helpless deer, blasting away different parts of the doe's body. These events seemed just as senseless. Several horrifying screams drew his attention to two young girls, in their early teens at most, their cloths ripped from their bodies. Their faces were cut and bleeding as numerous men, laughing happily, continuously raped them. Grendel's rapid heart beat and high blood pressure revealed his great agitation which was recorded in the chaotic activity of his brain waves. Gulping for air, Grendel could hardly breathe. Appalled, Parker found the footage unbearably disturbing, but forced himself not to turn away. He had to confront the truth, no matter how savage.

Nearby, screaming children fled toward the protection of the forest. An infant rolled on the ground, uncontrollably crying, its mother held down by one man while another barbarically abused her, performing acts too horrific to describe. Other men were rampaging through the small settlement, smashing some things while keeping anything of value. They were wild animals, more like uncontrollable savages than human beings, yet they were the fathers and husbands of Father John's flock. They were the products of civilization, good Christians all of them, embracing the traditions and customs that clearly made mankind superior to beasts.

After a few more minutes of unspeakable behaviors, brutal mutilations on living people, bloody carnage that surpassed anything known in the animal kingdom, the leader, who was the bearded stalker of Grendel in one of the earlier tapes, called to the men to assemble around him. Without a word, he held up over his head a large book that he had confiscated from one of the houses. Parker immediately recognized that it was The Koran. "This book calls us infidels and rejects Jesus Christ, our Lord and Savior." Suddenly, with all of his might, as if the book were something to be cursed and despised, he threw it into the flames of a nearby building. As its pages were consumed, its holy words blotted out in a sea of red, the men screamed their approval.

"Men," the leader said, "God is pleased with us for we have made His earth cleaner. The worshippers of Allah must be eliminated so we can enjoy God's pure love and the beautiful land that He has promised us," he said in a confident and reassuring voice, as if he might once have been a man of the cloth. "These nonbelievers are God's enemies, choosing not to glorify God's son, who was crucified for us. You have all read where it is written that God's enemies all must perish." The men nodded.

"We must complete the task so we can get home soon. God has spoken! We must dispatch the women and children so their vile blood can no longer infect the world with their evil lies." With those words spoken, he grabbed a whimpering infant from the road and quickly broke its neck as if snapping open a can of soda. To the bearded leader the small, helpless child could not have been human, could not have been like his children when they were born. With that depraved act the other men quickly spread out, shooting at point blank range and repeatedly stabbing the maimed and unarmed people "in the name of God."

Suddenly Grendel was in motion, as if a flood gate suddenly broke inside his head. His eyes were focused on two boys and a girl who were hiding behind a small shed on the outskirts of the village and not far from the dense underbrush of the woods. The children, none of them older than ten, were huddled together; the girl, the youngest of the three, was sobbing hysterically while sucking her thumb. One boy, eyes closed, held his hands over his ears, while the other boy, already traumatized, stared off into space. The militants were slowly making their way to the fringes of the village and soon would discover them. In no time, Grendel reached the children who were frightened out of their minds. He rapidly threw one of the boys over his shoulder and carrying the other children, one in each arm, he bounded into the woods.

With the sound of screams and gunfire in the background, Grendel quickly proceeded over a stream and moved deeper into the forest, his powerful body lifting the children over fallen trees and rocky outcrops as if they were leaves in the autumn breeze. Every so often, he would look down at the little girl who had a pink ribbon in her dark hair. She now had a heavenly smile on her face as if she were enjoying her first merry-go-round ride. Eventually he slowed down, and to further comfort the children, he started to hum a comforting melody coming from deep within. It almost sounded like a lullaby. The events happened so quickly that Parker figured that the four of them made a clean escape. At least, that is what he hoped.

Nauseated, Parker shut off the tape and looked with disbelief at Jan who was slowly shaking his head, stunned. There were many records of dogs saving children from fires or other life threatening disasters, but in most cases the dogs rescued someone that they knew and were attached to. Of course dolphins on occasion had rescued drowning swimmers, feats whose motives still remain a mystery. Parker concluded that some evolved mammals clearly had a reverence for life that at times was on par with humans. He knew that this was well documented among apes. "It looks like Grendel has a conscience and finds wanton violence painful to watch," Parker said, breaking the silence.

"He saved their lives," Jan said, still in disbelief.

Parker wondered, *would this tape be enough to convince the world of the validity of the Grendel Project or would critics still claim that the footage was bogus and the blood bath the product of masterful film making. Someone must be able to identify the burned down village and know what actually happened there. But what is the likelihood that a person would volunteer the information and risk his own life?* One thing was for certain, Parker realized that the children, who might still remember where they lived, must never see this sequence.

This was the unexpected, defining moment of the Grendel Project! It was the point where Grendel stepped over a mystical, spiritual line and behaved with a moral imperative, a level of consciousness far higher than any of the adult humans on the tape. Grendel's actions were those of someone possessing a soul. The more Parker thought about it, the less he felt surprised. His uncontaminated, intuitive energy just did what was natural. His actions were the basis for the highest expression, **not** following dogma as civilized man asserted, but yielding to his innate, untrained spirit. Left brain cultural indoctrination, even which claimed to be divinely inspired, could justify virtually any behavior, no matter how reprehensible. He had read hundreds of such accounts. The horrific events captured in the videotape were just another sad example of man's utter inhumanity to man, justified by some noble rationalization.

"Atrocities like that, maybe even worse, were common around here ten years ago," Jan said in a forlorn voice, his hand slowly running through his short hair.

"What makes it so difficult to understand is that the Christian militants were in no way directly threatened by those villagers. They probably didn't know any of them by name. They drove down from the mountain, raped and slaughtered them and then calmly drove home to their families... as if it were just another day of working the fields. They didn't claim their land. All of this senselessness was carried out over left brain concepts- the **idea** that these people were different because they practiced another religion-and that was the threat that produced this mayhem. Man is the only animal that commits unspeakable crimes based on nothing more than conceptualizations that are mentioned in their 'sacred' books. These Godly instructions support their totally distorted thinking, justifying a bloody massacre based on the mental calculations of a few villagers about how their future might be changed for better or worse. No animal, to my knowledge, is capable of doing this," Parker lamented in a totally exasperated voice.

"But the militants were in a war," Jan said, "to determine whether Foca was going to be Christian or Muslim. Their heritage, in their mind at least, their whole identity and the identity of their children, and their children's children was called into question. In a way this was an answer to the question of whose customs were going to prevail. It determined how people were going to dress? What they were going to eat? What holidays they were going to celebrate? What values they were going to follow? It is quite basic."

"And they could not settle that conflict peacefully?"

"These are old cultures. They are not very open to change," Jan responded, knowing that his excuse was weak.

"So now the problem no longer exists. Ironically, once again, you can go back to the Bible for perspective. Cain resolved his differences with his brother in the same manner. Because he did not have a blood sacrifice to 'properly' honor God as the good book demands, he killed his brother who did. Again, the same basic response over a conceptualization: whose way of worshiping God will prevail? Or who will have the most favor in His eyes? Animal behavior is a response to direct threats, mostly regarding mating and protecting their territory. Unfortunately, for this crazy animal called mankind, the **idea** of 'something'- whether it is the notion of a blood sacrifice or the teaching of different customs-is often **more** powerful than the threat of actual reality...like a person holding a gun to your head. We have a long history of our left brain resolving conflicts over religious "ideas" with unthinkable brutality or even genocide. The twenty first-century, as enlightened as we think it is, appears no different."

"I still don't believe what I just saw," Jan noted, continuing to slowly shake his head.

"It wouldn't take much, as heretical as it might sound, to make Grendel out as a Christ-like figure. He has had no religious training, yet his instincts are pure. Even though human beings are trying to kill him, he saves vulnerable children as an act of mercy, an act of unconditional love for all life, not just his kind. It is an excellent example of the New Testament injunction, 'love thy enemy as thyself.' He potentially sacrifices his life to correct what he sees as an obvious wrong committed by an overly aggressive and, at time, insane species. One, I might add, that is also trying to do away with him. It couldn't be clearer: Grendel, natural man, the avatar, the innocent, the highest consciousness on the videotape and a clear manifestation of pure light."

# CHAPTER TEN: THE AFTERMATH

After regaining his composure, Parker said, "Let's see what is left." The last sequence was short and very sketchy because it occurred in the dead of night, sometime after midnight. Fortunately, the camera's night vision worked perfectly, providing enough illumination to recognize that they were moving about in the downtown section of Foca. Grendel was softly humming. From time to time, he looked down at the slumbering children in his arms. When a car drove slowly down the main street, perhaps the police or just a person looking for somebody, Grendel ducked into a dim, narrow alley and waited. Once again his body tensed, ready for action. Parker felt apprehensive knowing that there was little time remaining. *How would it end?* When the car passed, Grendel was on the move again. For what seemed like the longest time, he ducked and weaved until he reached the city park directly across from the Foca's largest and oldest Orthodox Church. A floodlight focused on the ornate cross at the top of the dome. He paused and for over half a minute Grendel stared upward, as if the golden rood were all that still had meaning in the universe. For some unknown reason it appeared to provide comfort during his moment of confusion. He first was attracted to the cross on that brutally hot day when Father John led a pilgrimage up the mountain. Later he was mesmerized by it as he stared at the priest's headpiece during the funeral. As he rested, looking at the reassuring symbol, a sudden squall passed through the town and large flakes of snow started to fall from the blackness, dancing frantically in the light's shaft of brilliant illumination.

After a minute or two, Grendel crossed the street and sought the protection of a partially enclosed area in front of the massive church doors. For the moment they were safe from the wind and snow. He gently placed the children on a large wooden bench. The girl, the pink ribbon still firmly tied around her dark hair, suddenly opened her eyes. Perhaps contact with the cold seat startled her. She stared at Grendel's third eye which must have been slightly aglow and occasionally glittering from light reflected by the flood lamp. As Grendel slowly began to withdraw, her hand grabbed the mysteriously glimmering orb, ripping it from his forehead. Holding it close to her face, she studied it for a few seconds, and then she extended her hand as if she wanted to give it back.

Grendel was gone! She dropped the micro-sized camera, Parker could tell, because the transmissions ended with an eerie motionless image of a nearby stained glass window. The picture depicted Christ kneeling in the garden, praying, just hours before his crucifixion. The tapes concluded with that foreboding, motionless image, Christ looking upward for help that never arrived. Shortly after dawn the little girl was found safe, shivering on the wooden bench, while the boys were later discovered in the nearby park, huddled together and dazed. Everything suddenly made sense to Parker. The tape confirmed the seemingly preposterous stories of Alex, KA and the girl. It was **all** true. *But what about Grendel's fate?*

The following morning Jan drove Parker to Foca's massive cathedral. Parker was certain that the entranceway was the ending point of the tapes. Although he diligently scoured the ground, feverishly looking for what he thought of as a miniature diamond, he could not find the microchip camera. *What is the likelihood that it would survive after ten years?* There was no choice but to begin his search here, on the grounds of the old cathedral, and move backward in time.

They located Father Nikolei, the oldest priest in the parish. In a businesslike manner he confirmed that three Muslim children were mysteriously found in the area about ten years ago. The girl, he claimed, was adopted by a very loving woman, a member of his parish. He proudly boasted that the girl regularly attended mass. The boys, he thought, were probably put into a nearby orphanage, but he was unclear on that detail. He knew nothing about how the children arrived at the church nor did he know anything about the ultimate fate of the boys. When asked if they could contact the mother of the girl, he hesitated and then declined, giving no reason. He volunteered nothing more. After an awkward moment of silence, he excused himself, claiming that he had matters that required his attention. Parker's last hope was to find Father John and his small, village in the mountains where the transmissions began.

After several fruitless conversations, Jan's inquiries regarding Father John finally yielded directions to his church. They were in luck. A young priest remembered the name of the town-several unusual syllables that Parker struggled to pronounce-and showed Jan how to get there by sketching the route on the back of a church bulletin.

The hour ride, mostly on narrow, country roads, some paved, some not, went quickly because of the lively discussion. Jan, a psychology major at the university, had minored in philosophy. He confessed that he was an atheist, contending that there could not be an almighty God given the present condition of the world. The incredible amount of senseless violence, some of which he had personally witnessed, convinced him that there was no special protection for the innocent. He acknowledged that he was unclear on many of the church's positions. What little he knew made the church's doctrines appear to be illogical or, worse, blatant contradictions.

"If the human body-I have heard some people refer to it as the spirit's temple-is the house of the soul, does it not perform a sacred duty by staying alive? I mean food and sex are essential to the survival of our species. I don't understand why Christians so harshly condemn the flesh," Jan asked, raising his hand in protest.

"Walt Whitman, one of my favorite American poets and a man many people contend lived on a higher level of consciousness, maintained that the body and the soul are interconnected and that one is just as essential and holy as the other. The more unnatural the environment, Whitman thought, the more distorted the ensuing behaviors. Many holy men from various traditions agree and, like Whitman, propose a balance between the spirit and the flesh. But the church disagrees, claiming that the cause of the problem, the basic sinfulness of the flesh, is located not in the shortcomings of the environment, but in the innate depravity of man," Parker stated.

"So a pure God planted an evil seed within man?"

"Somehow the evil seed is attributed to Satan who continually exposes us to temptations. Because we are tainted, we inevitably fall. Our weakness, according to the doctrines, is our attachment to our senses, to our flesh. Yet, many other people, even some liberal theologians, claim that our senses are God-provided and therefore very sacred. For instance, the preparation and consumption of food or the sharing of a sexual encounter with one's beloved, when approached through higher levels of consciousness, these experiences suddenly can be seen as holy in nature. Ironically, the church even recognizes this with practices like Holy Communion and Holy Matrimony. Of course, when performed through lower consciousness, food and sex are simply gluttony and lust."

"That seems obvious enough, so what is the church's problem?"

"Most institutions have trouble finding a mid point or natural harmony because their dogma often leads to extremes. Every one of the five senses clearly is important and essential to our survival," Parker stated emphatically. "But the issue is moderation. When there is abuse, when sense gratification becomes obsessive and consumes a person's life, it's usually at the expense of the spirit. Hedonists glorify the abuse, believing you can't have too much good food or sex, while the church, quite justifiably, condemns it.

"That makes sense…so why not embrace man's innate goodness as well as his positive sensations, if there are such things," Jan asked?

"The religions of Abraham have struggled balancing the physical with the spiritual. Christianity primarily offers salvation, not so much for the body, but the soul. Even more problematic, the church portrays natural man as excessive in his appetites and indifferent to the promptings his conscience. Consequently their teachings denigrate the physical world as a whole. They do that because they have something to sell, the immortality of the soul… life everlasting! And the body, Jan, is not involved. According to them, the physical world offers only momentary pleasures and a lifetime of suffering, while the soul offers not only hope, but eternity."

"But what about nature? That's physical yet created by God."

"Some branches of Christianity even consider nature as tarnished and in some ways evil. For them the equation is hard fast and clear cut: molecules and atoms ultimately are sinful while the soul is holy. But the whole notion of balance, that the ideal person is a perfectly functioning body coupled with a pure spirit, that never seems to enter the picture."

"So many of the extremely, unnatural behaviors that we see all around us and, as some of my psychology professors claim pass for normal, you know, things like anger, repression and depression, these aren't triggered by our responses to our frantic and out-of-control lifestyles?"

"Nope, the church insists that man is innately sinful, end of discussion," Parker said as they rounded a sharp turn and headed up a steep gravel road with only large boulders protecting them from a thousand foot drop off.

"So we got that way because our natural state is inherently bad; we are tempted and we submit to Satan. That's it?"

"But the idea of a complete person, you know, one who experiences harmony between mind, body, soul and emotions, that idea doesn't make it into their dialogue," Parker said with a sigh. "Nor do they consider that because our lives are so fast-paced and complex, most of us in time get out of whack. We all experience profound alienation. The process of living distances us from ourselves, usually from most people, from nature and, as in your case, from a sense of the Source."

"You don't have to believe in a God to accept that some activities are better suited for a happy existence than others. I think that's the premise of Greek philosophy, isn't it," Jan asked?

"You are right. Balance, equilibrium, harmony, moderation, they all had a part to play for the Greeks. They realized that a healthy soul was much more likely to generate a healthy body. **Both** were mirrors of divinity. Our religions often seem preoccupied with negativity. Instead of 'Thou shall not,

Thou shall not,' a person should seek the highest as he or she understands it. A vibrant body surely would be part of that picture," Parker added.

"The Greeks not only encouraged excellence, but as I understand it, the complete person was part of their ideal…almost exactly as you describe it."

"What's interesting to me is that some eastern traditions, through different forms of yoga and T'ai Chi, actually encourage the development of the body and mind working together. Using their bodies to guide them, they provide ways for each person to directly connect with the divine energy. But the church doesn't seem to grasp the paradigm that body, mind and soul **are** interconnected, each a sacred part of a divine whole," Parker said, directing Jan to stop the car at a makeshift, roadside shrine. Several bouquets of mountain flowers were draped in front of a large, stone cross. *The work of father John's parish*, Parker thought as he admired the simple beauty of the marker.

As they drove on, Jan continued their discussion. "Maybe that requires too much discipline and too much time for the average person."

"Or maybe it's about the numbness of the soul," Parker noted. "When our being grasps that all of our life is holy, that's what could really change the quality of our lives."

"As an atheist I see excellent physical conditioning as part of the best life; however, it is a goal; like a Greek ideal, it is something to strive for. But it has absolutely nothing to do with a Creator, and for me, it's a stretch to couple something like yoga with divinity," Jan said matter of factly.

"The idea of God, of the most sacred, obviously includes only the purest and highest understanding of life. A person has a consciousness or inner being, even the most practical philosophies acknowledge that and I don't think that is debatable. You would agree, would you not?"

"Yes."

"Well take this very moment. Who is speaking? Who is listening? There is 'energy' within you and me that is engaged in the process of being alive. That center of being, that 'energy,' that is your real self. It's not a Greek idea, but very tangible. The *real* question, then, is whether that 'energy' is special, some would say holy, and connected to a larger energy grid, a Source if you will?"

"We can agree that there is 'energy' as you call it, but it's just energy. Universal principles, like gravity or space exist, but God isn't part of the process."

"That's because you experience life primarily through your mind. For you the energy is just vitality that resides within and is also responsible for running a living being. But, Jan, the crazy thing is that the mind isn't able to experience a unity with the absolute…that can only come from within, through a soul or inner being…through what some traditions call the heart.

I would like to propose a metaphor for you to think about. No matter where we are on earth, there are literally hundreds of radio waves that we could hear if we had a strong enough radio. But most people never hear the music because they don't know how to turn on the radio, to **open** their hearts. Jan, you have spent your whole life feeding your mind, but just what have you done to activate the radio inside of you?"

"I have never heard the music that you speak of…and I don't think I ever will," he said, dismissing the idea as if it were a fairy tale.

As the car rounded a corner, they entered the village frequently photographed in the tapes. It was exceptionally clean and thoroughly modest. The church, diminutive compared to Foca's grandiose cathedral, rested at the very center of the town and was flanked by a large cemetery on one side and a small garden with statues of various saints on the other. They parked and made their way up the old, stone steps, their sharp edges smoothed by centuries of wear.

Jan found the good Father in a basement office. Surprised by the intrusion, Father John hurriedly crushed his just lit cigarette in a nearby ashtray and kindly greeted the visitors. He smiled warmly, not knowing the men's intentions, as his eyes scrutinized first Jan and then the professor.

In real life the Father seemed somewhat shorter than he looked on TV. His voice was of a slightly higher pitch and he spoke nervously, as if he were exceedingly self-conscious. For a moment, Parker found him humorous, a diminutive man, dressed in a black garb, yet representing God's omnipotent authority on earth.

The conversation was awkward. Jan did his best to introduce Parker, but being an American professor of comparative religion hardly seemed an adequate justification for their presence in the basement of his church. Even finding a place to begin a dialogue was difficult. Why were they here? Unfortunately, Parker had not given enough thought to their meeting.

Parker smiled, trying to penetrate the cold formality. "I have seen your picture so often that I feel I know you."

"Have we met," the Father asked, his eyes still suspicious?

"No, not formally. But I have had a chance to watch you on a video tape. That's whywe are here. Perhaps you could help us."

"You have watched me in America?"

"Yes. It's a strange story… one that I want to ask you about." Parker paused, searching for a beginning point and then he quickly came to the point. "Are you familiar with a giant creature that lives in the mountains high above your town?"

After clearing his throat, Father John replied, "I'm not sure that I know what you are talking about. Giant creatures you say?"

"The one that you believed killed one of your townsmen about ten years ago and you presided at his funeral. Do you remember? You called the dead man a fallen hero."

"I remember his death," Father John said slowly, scratching his head. "But I am not clear on the circumstances...or this creature you mention."

"I have several news articles about the event. Would you like to see them?"

"No. I have a general recollection," he said, his voice suddenly sounding far away.

"I'm wondering what you can tell me about the creature."

"There have been lots of stories about it, but we are not sure if it actually existed."

"But in one article you call it evil and state that it must be killed. Why kill it?"

"The Bible, I'm sure you are familiar with that Mr. Professor," he said unsuccessfully restraining his sarcasm, "demands that all evil be wiped out. A giant creature is not part of God's plan. The creature is the work of Satan. Like a witch, or some other abnormality, it should not live."

"Other than that one death, which is questionable, is there other evidence of the creature's evil behavior?"

"You need no evidence," the Father said hastily, no longer posturing for the intruders. "Just its existence is evil. It is ungodly, like a witch. That is enough...it should not be permitted to live."

"Your job as you see it, then, is to act I guess you would say in place of God, to carry out His will," Parker asked calmly, as if stating a fact.

"To 'honor' the word of God," he said forcibly, with a slight smile.

"And you are absolutely certain exactly what is evil?"

"Yes, in this case. That damned thing surely was evil. You know that it killed one of our townsmen." After a long pause, again clearing his throat, he stated flatly, "To be honest with you, since you came for information about the creature, we killed the damned thing some time ago. I have a picture of it somewhere. It was truly massive, so we nailed the abomination to a tree, actually putting eight inch spikes through its wrists, like it was a vampire. They barely held him. It was quite a scene; its arms fully extended...it stretched out at least four meters." The Father paused, opened his desk drawer, and then after rooting around a bit, produced a photograph. "See for yourself," he said, not hiding his pride. "I took that with one of those new digital cameras. What do you think of the resolution?"

"Did you bury it," Parker asked?

"Naa...we left it up in the mountains to rot, dangling off of a big tree. One guy draped some flowers around its gigantic head as it drooped down toward the ground…I'm not sure why. Anyway, there is no point in trying to locate it. It was some time ago. I'm sure there is nothing left of it, but the damned thing was unbelievably large…really something to see."

Parker put his head down to hide his tears. In a muffled voice he asked, "How did it happen…I mean his death?"

"Well, it must have been a fairly intelligent creature because it was hard to track. The men set many traps, some really quite elaborate, but it never came close to any of them. However, it liked to swim every so often. Our sniper got lucky. He was all set up, gun and scope ready to go. He was monitoring a pool of water beneath a cascade, and suddenly the creature appeared. As it was floating on his back, the sniper nailed him with one shot at a hundred and fifty meters, right between the eyes." Father John pressed his index finger to his forehead, the exact spot of Grendel's third eye. "Pretty damn incredible if you ask me."

"Now your village is safer? Things somehow are better," Parker sadly asked with a note of disdain in his voice.

"One of Satan's abominations has been erased," the Father boasted.

"Quite possibly you killed the last of a nearly extinct species. There was much that mankind could have learned from it," Parker said dejectedly. After a long pause he asked, "Would you consider showing the world your photo? It could add much to man's knowledge."

"Nothing good could come out of it…only more people snooping around. Who knows, some fool might claim it was an endangered species or some such nonsense and then they might try to hunt down some of my parish."

"Did it ever occur to you that you could have coexisted with it?"

"If you coexist with evil, you eventually become evil. Surely, Mr. Professor, you are aware of that. The Bible's instructions are quite clear. God wants the Earth cleansed. You must have read, 'Do not suffer a witch, or any other abomination, to live.'"

"So killing in cold blood, in some cases defenseless animals, in other cases, defenseless people, that's a righteous, sacred act?"

"God hates the unclean," he said again, as if it were self evident.

"So everyone who is not of your belief would be better off dead?"

"They are all going to hell. This way they get there just a little quicker," a slightly sadistic smile curled around his thin lips.

Disgusted and getting up to leave, Parker said as he restrained his anger, "Thank you…I have no more questions."

"Since you are so interested in the creature, I have a gift for you, something retrieved from the creature, perhaps it will bring you luck." He took from his desk drawer a massive front tooth that was about three times the size of a large man's tooth. As he handed it to Parker, he said, "In this world, there are many things from which we need protection. You never can have too much good fortune."

The trip back to Foca was long and painful. Parker was devastated over the senseless slaughter while Jan's first reaction was a sigh of relief.

"When you talked about the killing of the defenseless, I thought you might make reference to the video. Given their attitude about the expendability of people, I thought that if they knew we had evidence of their butchery, we might not walk out of their in one piece."

"Kill everything that's not like you…in the name of God," Parker moaned, indifferent to Jan's concerns.

"Now you know why I'm an atheist; if that's religion, I don't want any part of it."

"Jan, what you just witnessed is so far removed from the Christianity that I know that it doesn't even deserve a comment."

"But it all seems for them to go back to some very primal instincts, like fear as it relates to their survival and prosperity. It looks like God has become a security blanket, a comfortable justification for acting on one's anxieties."

"In the worst of worlds, God becomes anything you want," Parker lamented.

"In fact, when you think about it, Eric, when the three Muslim kids glorified Grendel, turning him into a quasi God, a savior of the helpless, that was more understandable than what I just heard from the good priest. If God is what we believe has the power to transform our lives…then certainly Grendel, as bizarre as it may seem, acted as a hero, a literal savior for the children."

"And just like Christ, Grendel was crucified! Nailed up onto a big tree to be photographed and mocked, his tooth being used as a souvenir," Parker said, gently sliding his fingers over all that remained of what he hoped would change the world.

"Obliterated simply because he was seen as different and, of course, feared."

"And totally undeserving of his fate," Parker noted.

"I have always wondered about one aspect of Christianity…perhaps you can help me," Jan asked in a philosophical manner. "If Christ were married and had several young children, like your kids, too young to protect themselves, and if their lives were in danger, let's say someone was going to murder them in cold blood, like how Grendel was murdered, how would the highest level of consciousness, the most perfect person if you will, conduct himself? What I really wonder, would Christ intervene? Would he do things to protect his children or would the most enlightened act be to allow them to be slaughtered?" After a short pause, Jan added, "there is no doubt in my mind that we are wired to defend our offsprings, most mammals are, and so would intervention not be the ultimate moral action?"

"Interestingly, that particular moral dilemma, to my knowledge, is not covered in the Bible. But once I asked that very question to a highly respected minister and he responded with, 'What do you think?' I wanted to say a crazy person is going to kill your sons and daughters, and you could stop it. What would **you** do? But I didn't. I still probably would not have gotten an answer."

After a long pause, Parker continued, "So much of life isn't distinguishing between good and evil, that's generally fairly easy, but selecting the 'higher' moral position to live by. Personally, I think preserving innocence is the ultimate moral imperative. How could it not be? I suspect that Christian Gnostics wouldn't have much trouble with the question. They would look within and embrace what they experienced as the superior course of action. Their response might even vary from person to person, which would be a problem for most of the world's cultures because institutions prefer moral codes to be black and white." Parker paused and then added with sarcasm, "The church, however, probably would require some type of official decree or position statement on the preservation of the defenseless, linking it up to the Pope or some higher moral authority. Certainly an unschooled individual could never figure it out on his or her own...no matter how pure their spirit."

Parker's return to Foca was brief. With Jan once again translating, he paid the clerk for his hotel room; then they had a light meal at a nearby restaurant. For Parker much of the food that he had consumed over the past few days, like his whole Foca experience, was hard to swallow. They ate quietly, Parker picking at his meal, too depressed to converse. When finished, he thanked his assistant, warmly embraced him, and said goodbye. Jan went to the train station located in the business district; Parker started his long journey home.

As he retraced his trip back to the Belgrade airport, he felt the anger of profound disappointment. He had come seeking clarity and now, painfully, the mysteries were solved. His professional life, however, was another matter. Eric Parker, the controversial professor of comparative religion, the distinguished author of numerous books, the voice in the great dialogue over the nature of the sacred, no longer existed. Stripped of his identity and the expectations that were associated with maintaining a place in one's culture, Parker was returning, not just as a defeated man who had lost the dream of a lifetime, but, conversely, as a truly free man ready to embark on the ultimate quest. He wondered, *would it be wise to return to his family and put them in jeopardy? The challenge now,* he thought, *was to attempt to retain his anonymous persona, one seeking to live by the wisdom of intuitive traditions, in the complex and often absurd world of twenty-first century America.*

# CHAPTER ELEVEN: THE RETURN

As the plane touched down on home soil, he felt a sense of relief and at the same time a nagging anxiety. He drifted through the airport, happy to see familiar images and trite advertisements, as well as hear people chattering in English. Yet in America he was still a marked man. Nothing had changed in the five days that he was abroad. As long as he was Eric Parker, he was going to be stalked and hunted, just as Grendel was. He had offended the mullahs and he was required to pay with his blood, even perhaps with a bullet between his eyes. Unconsciously, he touched the scar on his forehead; it still felt lumpy and a bit sensitive. This reminder of the dark, inexplicable forces of chaos and destruction was always as close as a gentle glance of his fingertips.

Like Grendel, he had unintentionally disturbed the delicate equilibrium between progressive thinking and tradition; he had crossed an invisible line. He knew the truth, that a parallel species, intelligent and innately right brained, existed. But the world would have to wait to learn of it until the rigid teachings of the fundamentalists had lost power with the masses. Still, so many questions remained. *Did that species have a stronger avenue to the transcendent than mankind? Would life on this planet, in the great struggle for dominance that began thousands of years ago, be substantially different if Grendel's DNA, the genetic coding of the missing link, were to have prevailed rather than mankind's? If that had occurred, would the sacred be actually experienced and lived rather than mostly just talked about?* One could only conjecture.

The sad reality, however, was that left brained cultures always crushed right brained people. He remembered studies that analyzed the old reptilian brain, located at the base of man's skull. It was responsible for everything primitive, aggressive and blood thirsty in man's makeup. No doubt the old brain, as some scientists called it, was the bottom floor, the lowest level of man's consciousness. *Could it be that evolution had ceased and now mankind was in the process of devolving? Was the cold, calculating, greedy portion of the human mind systematically developed at the expense of our souls?* Parker thought that it was possible.

The Grendel Project offered a vision of mankind that might have helped usher in a new reality picture, one of compassion and harmony, one based not on the gratification of the "happy ego," but the true contentment of egolessness. Instead, with the ignorant assistance of time revered traditions, man was closer to self annihilation than ever. There was no shortage of fear, hatred, distrust and greed, all traits triggered by the over civilized left brain.

As he drove his rental car from the airport over familiar highways toward Durham University, Parker reviewed his hours of thought on the transatlantic flight. Unable to sleep, he searched for a way of life that refined and expanded man's capacity for a deeper level of awareness which he believed would create a fuller existence. For the world to move forward, traditional religion had to move aside, or at the very least, undergo substantial modification.

Parker embraced a simple answer: a new reality required man to create a new relationship to the sacred, and, perhaps, a new covenant. He clarified his thinking for his personal understanding only, with no intentions of creating a new creed for others to follow. There was no doctrine, just the simple belief that all genuine ways of life had to be eclectic. Each worshiper had to create "their way" from the myriad of sources that most resonated with their soul.

The first step of his ideal vision required that each person seek his most perfect self as he or she understood it. Then they had to have the courage to live by manifesting the highest levels of consciousness that they were capable of maintaining. Hopefully, they would directly experience the Source by opening their hearts and allowing the pure, universal energy to flood through, inspiring a new relationship with time and space. Each of us, according to our development and circumstances, had to radiate the intuitive, right brain qualities of universal love, compassion and harmony to the greatest extent possible. If an avatar like Christ or Buddha were useful in their devotion, each person was free to glorify the highest in a manner that most powerfully transformed their immediate, personal ground of reality.

Of course critics would claim that the ordinary person, regardless of the culture or religion, was unlikely to have a direct perception of divinity; therefore, consciousness transformation could never be a requirement for any belief system. At present, this was largely true. But Parker envisioned a new paradigm, one where each day, for about half an hour, people intensely meditated, prayed or utilized other forms of transcendence. The truth, Parker believed, was simple enough. Unless the soul, like a garden ready to receive seed, was constantly cultivated, there was no place for the sacred to enter. This was especially true in a culture that constantly desensitized the masses by the mindless and repetitive demands of ordinary life.

The whole purpose of religion needed to be re-examined. Parker wondered, *was the seeking of divinity primarily a way to remove people from their personal*

*misery through the notion of salvation with its assertion of life everlasting, or
was the quest intended to reveal to each of us our highest human possibilities,
those traits that allowed us to move towards egolessness and transcendence?* Year
after year Parker had witnessed society's agenda playing out in his classroom.
The majority of his students wanted a high grade, his culture's symbol for
excellence. Yet they had no intention of grasping, on the deepest levels, the
intricacies of world religions. College was only another performance. For
them, the course always remained safely outside of their inner being, beyond
their personal reality picture. They took notes, wrote papers, passed exams, and
in some cases truly excelled, but most experienced no inner transformation.
Only a very few internalized the process, personally encountering the ideas,
and then actualizing the ones that truly resonated with their souls.

His other realization was that the religion of the future would best resemble
Zen Buddhism in that it would be an **experience rather than a system of
thought**. There would be no need for standardized dogma, no essential
rituals to perform, and no old scriptures, translated from various languages,
to memorize and then passively embrace. A relevant religion needed to give
priority to the present rather than hundreds of pages of history. Certainly it
was not necessary to have a hierarchy of leaders, giving certain people more
power than others and creating unnecessary competition for control.

Furthermore, there would be no official moral code that mandated
people's behavior. Right conduct would be determined by the individual as
an outgrowth of their level of consciousness in relationship to their unique
situation. Each person would take responsibility for their actions, deciding
things like when aggressiveness was appropriate or whether sexual intimacy
was a response to lust or love. The ability to live a truly sacred life required
much more than simple submission to another's view of an elevated existence.
If a person's life were filled with avoidable suffering, then they, alone, would
be required to make new choices based on what they learned from their
mistakes. Each person would take an active part in shaping a meaningful
moment.

Our relationship to the external world would be a manifestation of our
understanding that everything and everybody were interconnected. Each
person would be perceived as an embodiment of the **same** sacred energy that
one encountered in themselves, yet outwardly, we would all be different. The
natural world would not be perceived as a set of savage forces, blind energies,
but a multitude of universal principles that were part of a cosmic dance that
reflected a divine harmony, a unity that the individual did not necessarily
totally perceive or fully understand.

The elevated person would approach nature, which he would have no
need to control, with the Buddhist maxim: "**First, do no harm.**" Even the

cold, dark expanse called outer space, often seen as absolutely lifeless, would be understood in a new manner, as part of a universal rhythm, a mysterious process that only man's soul could grasp. There would be no need for personal resurrection, no justification for individual salvation; each of us in the here and now would react to the external world through a **sacred** connection to the other. Ultimately, with purified consciousness we would see nothing but God. We would live in a pure land that perfectly reflected the flow of the universe.

Of course it was incredibly idealistic to think that western cultures no longer would be dominated by an obsession for possessions. For a brief period it did occur. Parker was too young to personally recall "The Sixties," but his readings suggested that large numbers of educated people for a few years in the twentieth century lived a value system that rejected materialism in favor of authenticity, unconditional love and universal peace. The hippies fully embraced the lyrics of John Lennon's hit single, "Imagine," truly dreaming of a world of individual freedom and acceptance.

If enough people were to prioritize the genuine and the transcendent, civilization could shift from unrestrained consumption to harmony and cooperation, the only qualities capable of saving the planet. The new reality that Parker constructed in his head had to be like the hippie movement, an attitude that was not limited by race, religion or national affiliation. It could only be transmitted by the popular culture. Once again, the young across the world would be the first to grasp the magnitude and necessity of a new consciousness. Imagine, each person living through the light of their soul, allowing the joyful, natural energy within them to connect with other beings until all the world was unified into a radiant dance, a sacred, glorious synchronicity.

Because he didn't know how his return would be received, Parker decided to approach the university through little-used back avenues. He felt uncomfortable trying to remain anonymous, but what he wanted to share with Bronsky was best communicated in person. He turned left onto High Street, cruised past the royal blue Durham University sign with its ironic slogan, "Truth is the Foundation for the Highest Life." Then he drove up the broad thoroughfare, along the ten foot wide center section which separated the road, passing now mostly bare trees and mounds of dirt where flowers until recently were planted. He parked in the faculty lot, and hurried towards the side entrance which provided an inconspicuous access to both the classrooms and the professors' offices. He swiped his plastic access card; the door opened.

Bronsky deserved to know Grendel's fate. He hoped that the tooth, the only physical evidence of Grendel's existence, might be useful as well. If scientists were to run a DNA scan on it, they might be able to confirm the existence of a parallel species. She still was connected to friends who regularly worked in the labs. Given the right circumstances, everything could be carried out in a most discreet manner. It was a long shot, but it was his only hope for vindication. If the DNA proved the existence of a new species, he would regain his credibility and reputation, qualities, however, that no longer seemed very valuable to him. Most importantly, he wanted to thank her for directing Jan, whose assistance was indispensable, to meet with him.

Parker quickly scurried up the corridor, finding the door to Bronsky's office unlocked. He sat on a large, leather couch, hoping that she would soon return, as was her custom, after her noon class. On an end table atop a pile of magazines, many of them professional journals, was the most recent edition of the school newspaper.

He was startled to see a large picture of himself on the cover, accompanied by the headline, "REINSTATE PARKER." The article that followed strongly defended his actions. It presented a clear account of the earlier harassment and the recent attempt on his life, including police reports and several interviews with professors who held him in high regard. The article read in part: "Parker was forced to go incognito not because of any wrongdoing regarding the Grendel Project, as the university alleges, but because his life and the welfare of his family were in jeopardy. President Knight overstepped his authority when he dismissed him without due process. The professor has the right to be heard in public. Then the facts can be examined. The university's witch hunt, reminiscent of the Puritan era, is an embarrassment to all enlightened people. Parker deserves protection and support, not condemnation, until this matter can be resolved in a rational manner. The only sensible course of action is to reinstate Parker." There was also information regarding a planned march protesting his uncalled-for firing. It was to proceed that evening from the School of Religion to the president's office.

For a moment he thought about returning and assuming his old duties, but he could not function always worrying about the gorilla-faced assassin. He could not live freely always looking over his shoulder, and constantly wondering about the safety of Kathy and the boys. No, his career was history.

As he further skimmed the paper, he found in the middle an editorial written by one of Bronsky's students. The point of view was particularly interesting. "The true purpose of education is consciousness transformation. Our society is in dire need of reformation. The university must accept its role as an agency for social change. In the pursuit for a better way of life, there can be no justification for Durham University's reactionary behavior and its willingness to bow to the scare tactics of pressure groups. No meaningful change can occur until the minds of students can be turned away from the forty-four gallons of beer that they consume each year and re-centered on the pressing issues of our day. The university must do more than simply provide reams of information; it must demand a stimulating learning environment that directly engages the student's deepest self. A student's mental faculty is grossly misused if he or she is only required to assimilate data. The process is pointless if it fails to have appropriate corresponding internal development. Genuine people seeking the best life possible are our planet's only hope. For real change to occur, we must develop as human beings. Rather than using the computer as a model, we must pattern our lives after those who have demonstrated the willingness to expand their awareness, not just their knowledge base. Consciousness transformation is all that can save America and can go a long way towards saving the world." Parker smiled to himself and wondered how many students really understood the problem and cared enough to find a solution.

Another essay from the op-ed section caught his eye. Everyone seemed to have an opinion about him. The headline, "PARKER'S FUTURE: NO RETURN" condemned his style of teaching as well as his attitude towards traditional religions. "Eric Parker, the so called professor of Comparative Religion, claims to encourage young students to think for themselves, however, his narrow mindedness towards traditional Christianity has damaged hundreds of students' belief systems. Parker contends that he is presenting other ways of thinking that are designed to encourage the student to challenge years of indoctrination by the church. Why not challenge other ways of thinking and leave the church alone?"

"Everywhere in secular society Christianity is under attack. Clerks in department stores have been instructed not to wish people a 'Merry Christmas.' Schools have banned the singing of traditional carols and have changed the name of "Christmas Recess" to "Winter Break." Students should not be required to take courses that undermine their faith. Parker stepped over the line with his so called Grendel 'research.' To suggest that a primitive, uneducated primate, if it actually exists at all, might be imbued with qualities that imply a religious consciousness is not only absurd but unconscionable. As children, we learn about Christ and the glories of salvation. What does this mutant know? Parker's suggestion that the sacred might be innate and part of our birthright is blatantly contradicted by the evil world we live in. Corruption is rampant. Sinfulness is man's natural condition. Durham University is better off without the controversial professor. From what I have witnessed, his expulsion is justified. Rather, he should spend his time reflecting on the nature of damnation." He noticed that it was penned by Erin, the freckle faced sophomore who criticized his free thinking. He skimmed a few more pages of the newspaper wondering what else awaited him.

As he started reading an interview of President Knight, his ears picked up a peculiar sound, a metallic tingling that he had never heard inside the building. At first, he was unsure what it was. When he realized that the sound was being made by a thick chain, he wondered why the custodian would be locking the main entrance in the middle of the day. He got up, glanced through Bronsky's office window, and without entering the hallway, located the source of the noise. His heart nearly jumped out of his mouth! The gorilla masked assailant, with two automatic weapons slung over his shoulder and several handguns around his waist, was in the process of padlocking the main door of the lecture hall. He knew that Dialing 9-1-1 would be useless. Help would arrive much too late. At that moment the universe was reduced to this dim hallway and his actions or inaction.

Parker's brain was flooded with gruesome thoughts and images. This had an eerie similarity to a recent slaughter at Virginia Tech where a deeply alienated young man killed thirty-two people because he was envious of their lifestyle. *But to kill in the name of religion, that was not just misguided, it was pure lunacy!*

His mind slipped back to the massacre at Foca where Father John's militia killed in cold blood people whose only crime was to have a different religion. Now his colleague and her students were at risk simply because she supported the Grendel Project. For a second, he saw Bronsky's classroom as a scene from a cheaply made horror movie. It was unbearable! Pools of blood were everywhere while young bodies were riddled with bullets and draped helter-skelter over desks and chairs. Others were wounded and helpless, trying to flee. Bronsky was decapitated, her bloody head resting on the podium at the front of the room as if it were a gruesome bust in a wax museum.

An uncontrollable rage came over him. There was no need to weigh the virtues of a moral code or consider the teachings of the holy ones regarding violence. "First, do no harm" did not apply to this situation. All he knew was that he could not permit this atrocity to take place and still be able to live with himself in the years to come. Perhaps his actions were more designed to save himself from a lifetime of anguish than rescue Bronsky and her room full of innocent students. A true motive was not clear. At that moment something snapped inside of his brain. He moved into another zone, beyond social conditioning and traditional moral values to a deeper level of being, to the promptings of his most basic energy.

While the assailant had his back to him, securing the chains and clicking the locks shut, Parker crouched beneath the door's lower wooden panel, gently leaning on it and twisting very slowly the handle until the lock released. As the deranged zealot walked towards Bronsky's classroom, Parker listened to his boots as they clunked heavily down the hallway. He remembered the boots of the militia, crushing clumps of autumn flowers, as they walked through the meadow on their way to the Muslim village. Time and movement stalled and almost stopped. He waited for what seemed like an eternity, his body as tight as a coiled spring, until the gunman was slightly beyond the door. Without thinking, his adrenaline surging as never before, he exploded into the hall, ramming the surprised assailant in the small of the back with such force that he fell heavily against the tile wall, smashing his head. Unable to use his hands to break the fall, he landed awkwardly on the floor. Stunned and disoriented, the gorilla-masked man never knew what actually hit him.

As a young man, Parker was never involved in contact sports, although he often ran five miles at a time to stay fit. Also, he never had any instruction in the manly art of self defense, so he was totally unaccustomed to physical confrontation. Although not a pacifist, he was uncomfortable with violence. Nevertheless, he found that his body was suddenly moving without his mind's directions. Was his action a primitive, blood thirsty urge to kill or was it a sacred power emanating from a desire to protect the innocent? He was in another world, one that he could not control. For that matter he may as well have been in another body. He was no longer Eric Parker but a ball of throbbing energy seeking to blot out a savage menace. Perhaps it was what Grendel had experienced when he saved the children. It was not premeditated, or for that matter, anything that he could have ever predicted. His body simply had a force of its own, regardless of what his brain might have been telling him.

In a flash Parker was on him, sitting on his chest, his knees pinning each arm to the floor so he could not use his weapons. He found his hands around the assailant's neck, pressing on his throat as tightly as possible. He was not thinking, *I am going to kill him,* for thought had abandoned him the moment he exploded from Bronsky's office. This embodiment of evil, this anonymous force of darkness, had to be obliterated. In the great scheme of things it mattered very little that a professor and her classroom of students, oblivious to the situation, were spared from certain oblivion. But for that one shining second Parker could make things right. The intruder offered very little resistance. Perhaps he was already unconscious from his head bouncing on the floor. Nonetheless, Parker's hands were like an unrelenting grip, constantly tightening on the forces of evil, choking not just the gorilla masked man, but Grendel's murders, Father John's rampaging militia and all of the darkness of misguided ignorance.

Then, time shot forward, jolting him from his surrealistic nightmare like a released clutch on a runaway truck. Suddenly his senses returned.

Astonished to find his fingers around a man's neck, his knees digging into human flesh, Parker did not immediately understand how he came to be on the floor. Shocked and still somewhat dazed, he slowly began to assess the truth. The lump that he was sitting on was no longer moving. He had just killed a human being. He let out a short grunt of surprise and then stood over the body, not as a victor but simply in a painful moment of realization. Right or wrong, Eric Parker, religion professor, had taken another man's life. His brain started to race as if already in flight, fleeing the repulsive reality that lay beneath him. There was no time to lift the mask and dispel the illusion. Quick, decisive action was the only thing that could save him.

The hallway was totally empty. It appeared that no one heard anything unusual, for there was no one peering out from any of the office doors. Soon the classes would conclude and dozens of students would energetically rush into the corridor. Parker sprinted down the hall and left through the little used side door that he entered. No doubt the dead, masked-faced body on the floor near Bronsky's classroom would cause a great commotion, especially since he was heavily armed and the main doorway was chained. No doubt the police would be called, but no one ever had to know the truth. Like Grendel, Parker wanted no acclaim for his heroic actions.

As he understood the moment, he simply allowed his natural energy, his innate sense of justice, to reach a fitting fulfillment. He was not acting on an inflated ideology or a warped sense of goodness. The situation could be reduced to this: one life was blotted out, sacrificed for the good of many. Hopefully he would never have to justify his actions in a court of law which sometimes seemed to have undeserved mercy on those who were at fault. He knew of a person jailed for killing an intruder who was going to do harm to his family. No telling how lawyers could twist things. Perhaps he could be accused of murder and condemned for using excessive force. *Yet who,* he wondered, *was in a position to unbiasly determine the most judicious course of action? What if the assailant regained consciousness and killed everyone? Then what would be deemed as appropriate? So many variables, so many possible scripts!*

He retraced his earlier steps, forcing himself to slow down and not appear as if he were running from something. More people were walking around the campus, some perhaps returning from lunch. Fortunately no one seemed to recognize him. He got into his car and quickly exited the university. Even though his mind was distracted, flooded with a jumble of conflicting emotions, he still managed to find his way to the expressway. Still numb and bewildered, he drove to nowhere in particular.

Parker knew that this small victory was not the end of his predicament. If he were to resume public life again, it would only be a matter of time before he would be discovered and his life would be exposed to the constant threat of death. Kathy and the boys were totally innocent, but their lives could be snuffed out in a heartbeat. The risk was too great. He loved them and always would, but his love for them required drastic changes. His future would have to take a new direction. He was headed northwest towards the Blue Ridge Mountains, towards Kathy and the boys, but as much as he would like to reunite with them, he knew that this was not the right time. Everything was in a state of confusion. There were so many questions that he personally had to resolve. If he couldn't be Eric Parker, an identity that he played well and rather liked, who could he be? **In fact, did he have to be anyone at all?**

Like Grendel with the children in his arms, Parker was in full flight, running from the darkness that seemed so overwhelming. He needed to cleanse himself, wash away the stench of an overly civilized society that produced hoards of sweet smelling bodies that reveled in their impurities. He, like so many others, lived a compromised existence. Sadly, for one to survive, his culture required a damaged value system, one that empowered prestige and wealth over authenticity and genuine connection. Yet, that was the world to which Kathy was attracted. He knew that there was no way that he could dislodge her from middle class America with its obsession with convenience and excess.

In fact, it was just a matter of time before Kathy would become a soccer mom and live in the senselessly cluttered world that he found so meaningless and unbearable. He knew that his family would be far more secure, although less prosperous, with him out of the picture. He deeply missed them and felt guilty about the possibility that the boys might grow up without a father, but battered parts of his spiritual self begged for completion. If he were going to burn away the impurities of his life and return to a state of childlike innocence, to what the Taoists called the uncarved block, he had to travel that long journey entirely alone. *Was it selfish to seek The Highest?*

As he continued to drive west, darkness slowly descended on the landscape. His thoughts, like dozens of bats returning to a cave, settled on the many failures of the American way of life. *From the beginning, he was required to perform. He learned that if he worked hard enough, he could earn enough money to allow him to acquire a suitable woman, maybe the angel in his dreams. Conversely, young females were brainwashed to believe that their primary task in this world was to become a man's fantasy woman by maximizing their physical beauty and their seductiveness. All around him life seemed nothing more than a series of impersonal and usually foolish transactions that people unconsciously carried out.* He deeply resented the charade.

As the hours passed, the miles piled up, pushing his thoughts still deeper. *He realized that he had been cleverly manipulated to believe that he was genuinely choosing his life but the truth was that he was carrying out society's blueprint as if it were imprinted on his soul. A lifetime of advertisements, about a quarter of a million a year, and millions of images from television were enough to distort everyone's perceptions. Now, all the game playing hardly seemed worth the effort. To him, the superficiality of the American way of life was an unbearable reality cage, just another form of death.*

He stopped for the night at a motel in Tennessee. It was a bit run down, needing a fresh coat of paint and some remodeling, but it was inexpensive and not far from the highway. Parker was still struggling with the violent events earlier in the day. He assured himself that there were many cases of Godly people standing up against evil. He thought of Dietrich Bonhoeffer, the German theologian who plotted to assassinate Adolph Hitler. His intervention, shortly before the war concluded, cost him his life. Yet Bonhoeffer's conscience demanded him to engage in what he considered to be a just and moral action. Parker knew that there were thousands of others just as courageous.

As he opened the door to his room, he remembered that Moses once took the life of a man, an Egyptian who was beating one of the Israelites. Apparently there were no other Egyptians in sight, so Moses killed him and buried him in the sand. The account sounded both premeditated and cold blooded. Yet curiously the Bible did not refer to Moses' actions as murder. Probably he felt just as justified in his response to the situation as Parker, who still felt unsettled and uncomfortable with himself, as if he were covered by some invisible grime. He had violated a cosmic law, but only as a last resort. He wondered *if an Egyptian beat to death an Israelite who was smiting an Egyptian, would the Bible call that murder?* Real moral standards could not be as arbitrary as these seemed. The truth was not about splitting hairs over how words, like "kill" or "murder," might be interpreted. As he drifted off to sleep, he asked, *how can it be wrong to intercept something that is unquestionably evil, and violently destroy it, before it has a chance to blot out innocence?*

The next morning was overcast and dreary with the threat of an ice storm. As Parker looked outside his window, he saw a sad looking older couple slowly walking toward the motel's restaurant, a dimly lit building attached to the office. They did not speak to each other. *He wondered how many cups of coffee they had drunk in their lives? Did the caffeine really make a difference? Were they any happier?* The few cars on the road seemed to be traveling in slow motion. *Was it caution caused by the weather or simply because they had no where important to go?*

The world seemed bleak and in the light of day, his accommodations looked especially shabby: old, musty drapes, a heavily stained rug and a broken TV. *Was everyone sleepwalking through a re-occurring dream that they called their life?* The ordinariness that typified most people's existences was unbearable to him. Twenty-seven thousand days, a lifetime, locked into social roles and routines so everyone could keep their place in the gigantic charade. It hardly seemed worth it! *Why was there such an obsession over consumption and conveniences, as if one's life were going to be changed in some really significant manner with an upgrade or new purchase?* What passed for life seemed so stale, like eating over and over again the same old, tasteless cake that decades ago seemed to offer some nourishment? He decided that the road could wait.

Parker did what seemed natural and satisfying. He sat cross legged on the floor, his back perfectly straight, his butt on a pillow from the bed, and began deep breathing and reciting his mantra in his head. Over and over again, in unison with his breathing, *Dear Lord Jesus, please have mercy upon us.* Slowly he drifted into a world of glorious light. A glowing, orange sun rose in the middle of his forehead, just like Grendel's magic eye. When he closed his eyes, the vibrant energy was always there, radiating out into the darkness of his inner cosmos. Like the sound of the sea in a large shell, the universal light eternally shone within him. He floated upward into radiant effulgence, what non- meditators might think was a drug-induced trance. Yet there was nothing artificial or unnatural about his elevation. For an indeterminate amount of time Parker was lifted up, blissfully floating in a state of total freedom, enjoying the nectar of the Gods. Sometimes he wondered if that was what our souls were destined to experience when we died. *Would he float in a continuous state of ecstasy?* Nobody can spend twenty-four hours a day in constant transcendence, but early each morning, journeying upward into paradise was a wonderful way of entering the day. As he slowly came back to his persona, a man on a quest for a truly meaningful life, he felt ready to attend to his future.

The energies that moved Parker were beyond the immediacy of his drab surroundings and the sunless sky. At the check out counter, an overweight, middle-aged woman greeted him with a flat "Good Morning." Then she processed his bill. Her lifeless eyes were fixed on a small TV set as she mechanically swiped his credit card. Without looking at him, she returned his card with the receipt, which he quickly signed, and then she said with complete indifference, having never once actually looked at him, "Have a nice day."

Outside, a pale faced man in his late forties smoked a cigarette, blankly staring at nothing. He reeked with the rancid smell of alcohol and vomit. The world seemed frozen, a lifeless montage of dead souls. When he walked towards his car, he was surprised to see a solitary white flower, a daisy, poking its head towards the overcast heavens. Out of season, somehow it had survived the earlier frosts. An old wooden birdhouse was attached to a nearby tree.

Parker wondered, *did it ever contain a nest or was it always an empty shell, like the run down motel, dangling on the edge of oblivion.*

When he continued driving, he found comfort in knowing that there was a permanent reality far superior to the transience of the tarnished physical world to which most people seemed so attached. Nature was beautiful but there was something far more powerful than mountains and trees. He thought of the great Hindu work, *The Bhagavad-Gita,* and its teachings on impermanence. Several lines came to mind: "He who kills or thinks they are killed, know not my ways." Parker felt a surge of reassurance knowing that nothing really ended. Everything was in the eternal process of returning. The spirits of Grendel and the gorilla masked man were imperishable and, in their own manner, would find their way to their next incarnation.

Suddenly it occurred to Parker that he wasn't running away from the relentless shadow of evil as much as he was beginning to return to where he truly belonged. Just one day removed from his life shattering drama and already the metaphorical light seemed brighter and his many painful burdens more manageable. He was no longer the proverbial caterpillar condemned only to a physical existence of eating leaves, sleeping and getting fat. Already he was beginning to transform, to form a mysterious spiritual cocoon. He hoped that soon he would become the glorious butterfly, afloat in the timeless meadows of eternity.

Shortly before noon, on a nondescript road in the middle of what seemed to be nowhere, he had an epiphany. He remembered an old Buddhist story about a monk who went for a walk and strayed too far from his temple. Suddenly a hungry tiger was stalking him. Soon he could flee no further for a cliff with a thousand foot drop blocked his way. As the beast neared, in a desperate attempt to escape, the monk shimmied down some vines growing on the face of the mountain. As he held on for dear life, he could faintly hear the vines cracking, ever so slowly breaking. With the blood-thirsty mouth of the tiger above and sure death on the rocks below, he looked from side to side, hoping for a miracle. Then he discovered it: **the perfect strawberry**. It was large, ripe and incredibly succulent. He plucked it and placed it in his mouth. It tasted delicious. *Yes,* he happily thought, remembering a most basic principle: *Be, here, now, for there is no sweeter moment than this moment.*

He realized that at that very moment he was grasping the true nature of his lifelong dream. There were no people, or stores or homes in sight, only desolate patches of woods with leafless trees and rolling fields with broken, pale yellow stalks. By all accounts he was surrounded by nothing of significance and utterly alone. Over the next rise in the road, more of the same: vast expanses of emptiness. This moment, in the midst of pure solitude and apparently void of any meaning, was the very paradigm that he was seeking, the ultimate strawberry, the truth he believed would, at last, liberate him.  He wanted his future to transport him into the blissfulness of total nothingness where he had no personal identity to manage, no ego to nourish, no audience to please, in fact, no self at all.  He hoped that he could learn to surrender, experiencing life as a vibrant energy spontaneously interacting in a total state of mindfulness with the immense power of a continuous now.

He remembered the statue of Lao Tzu, the founder of Taoism, which he once saw in a museum.  The master was standing on a pedestal, his hands peacefully folded over his robe.  Yet his head was neatly severed from his body and peacefully resting at his feet.  At the time, it seemed not just peculiar but quite startling…a headless Lao Tzu.  But now he understood not just the virtue, but the absolute necessity, of having no intellect, no brain that constantly engaged in continuous, useless mind chatter, as if it were a monkey in a jungle of perplexing thoughts.  He trusted that he could soon find his way to the radiance of this headless paradise!

A strange realization popped into his mind.  He remembered a book about people who literally died; their hearts stopped beating for ten minutes or less.  When their hearts began to beat again and they regained full consciousness, dozens of people, all having been pronounced medically "dead," recounted the same story.  They all encountered a beautiful, peaceful, loving light that drew their spirits toward it.  They each felt a sense of bliss that they never knew in life and each person wanted to remain in that warm glow forever. *Was this the very same light that he experienced each morning in meditation as his liberated soul momentarily took flight, letting go of his ego?  Was that the longed for TRUTH on the other side of the veil of illusion?  Were being headless, eating the perfect strawberry, basking in the light of pure effulgence, all varied faces of the sacred?*  He smiled, the long, continuous grin of momentary release.

# Chapter Twelve: Two Years Later, Almost Invisible

To the left brain, life is a series of paradoxes. When we appear to "solve" one seeming contradiction, others invariably emerge. Conflict is the mostly indigestible food of the constantly thinking mind. Escape requires us to be largely transcendent. At that stage the brain confronts only the most basic concerns, things like, "I am tired," or "I am hungry," important information needed to sustain the welfare of the body that houses the spirit. Thinking, then, becomes an auxiliary and very minor process in a soul directed existence. The brain stops generating problems to empower itself, and in time becomes a useful tool for the fulfillment of the inner being. When freed from endless monkey chatter, life simply becomes the awesome, natural energy of pure being, dancing unfettered in continuous moments of radiant light. We largely return to the effulgence of the beginning, the uncarved block, the permanent world beyond a person's little, mostly manufactured dramas.

In a few short years, Eric Parker, the popular professor of comparative religions, was able to almost fully disappear. It wasn't an act of will power orchestrated by the rational mind and based on instructions like, "I must keep a low profile" or "I must become one with my true self." He hardly noticed it, but over time a door opened wide within the deepest part of his being and light automatically surged forth, filling his whole life with clarity and love. Without any effort his being became buoyant with a joyful, irrepressible energy. Everything around him had the continuous glow of divinity.

Not only did he manage to shed his old name and erase all of his links with his past life, but more importantly, he was no longer the person whom society constructed and maintained for the better part of forty years. He let go of his career, his family and his identity. He realized, even with all of his good intentions, he was largely oblivious to the natural world that surrounded him. Over the course of his life he had lost the ability to see as a child. Even when he did "look," he rarely **saw** the moon and the stars, the fluffy clouds floating in a clear blue sky, or the tree play of birds and squirrels. His brain as well as his senses had become dull, blunted by years of senseless routine. Over decades and without ever being consciously aware of it, he had turned the present into a series of useless conceptualizations, much like he had with the past and the future.

For most of his life he had been oblivious, except on the most superficial level, to the lives of others, even those whom he had thought that he loved. Perhaps Kathy was right; maybe relationships were difficult for him. He had wanted total intimacy, but his brain with its constant "to do" list made that almost impossible. And for the most part, he knew nothing of the immensity of his own inner being. Without really knowing it, he had walled off significant parts of himself. Although he once thought that he lived in a spiritual castle, he realized that he was confined to a small outer room that appeared to lack both windows and doors.

The hoped for transformation, the longed for journey toward the highest, did not occur with regular progression, like a child in school moving from grade level to grade level. Much of what happened went totally unnoticed. Over time and with deeper and deeper meditation, he was no longer completely attached to the constant mental preoccupations that once dominated his life. There was no longer an incessant glaze of thought keeping his being from fully grasping and embracing his environment. There were more and more moments when thought somehow ceased and he simply witnessed life as it truly unfolded. The miracle of total mindfulness was incomprehensible.

Initially, he searched for a community that resembled the loving fellowship described by devout Christians in the Book of Acts. He yearned for a world of no possessions, no masks and no power struggles. But that was totally unrealistic. Eventually he found a very small, three room cottage on the outer fringes of a mountain village and over time slowly settled in. The proceeds of a modest inheritance from his recently deceased parents allowed him to buy outright the property that was sold as a "handy man special." His taxes were less than three hundred dollars a year. The thirty-five thousand dollars that remained, his rainy day fund, he hid in the house, preferring not to deal with a bank.

As a matter of practicality, he worked six or seven weeks a year doing manual labor: picking fruits and vegetables, repairing fences, selling firewood or assisting a carpenter. He enjoyed those jobs, finding the physical labor cleansing. His meager income, however, covered his simple lifestyle. A bicycle was his major form of transportation. Much of the food he ate, mostly vegetables, was primarily provided by his large garden. He had no computer, no telephone and the bare minimum of appliances, so his electric bill was insubstantial. In the winter he warmed his house with a wood stove. Because he enjoyed cutting and splitting the numerous downed trees in the nearby forest, he had a large woodpile. Sometimes he bartered with the cords that he did not need. He purchased an occasional article of clothing only when absolutely necessary, gladly sewing patches on the elbows of his old shirts and the knees of his worn pants. The Mountain Thrift Shop, which was only two miles away, provided suitable products for most of what he needed to replace at a fraction of the original price.

The only luxury that he permitted himself was an occasional paperback book. His new personal library omitted knowledge based, informative works, but rather focused on books that detailed the specific steps that a person needed to move towards personal transcendence and inner wisdom. He enjoyed reading accounts by the Dalai Lama, Thoreau and Gandhi, their journeys empowering his path.

He was happy to give up the accumulation of unneeded possessions; for that, he was more than compensated with a feeling of joyful liberation. Now, his life was devoted to expanding his level of self realization and to cultivating, as best as he could, a working Christ consciousness. He experienced the mystical energy that composed him and all other living creatures as hundreds of invisible, inner gates, each waiting to be opened, each providing a passage way to a fuller existence. Life, at its best, was a beautiful inner garden of sacred living, bathed in the joy of childlike firstness. The real point of it all, when he was finally able to actualize it, was to experience the sweetness of divinity through a new relationship to the universe, focusing on the perpetual light of infinite wisdom and infinite love.

A new understanding grew within him. He was aware that the cosmos was not an inanimate series of mechanical events as the old Newtonian model proposed. It was not something to be acted upon as one might tinker with a car engine or the workings of an old watch. Most certainly, it was not a series of scientific problems to be solved. The universe, which he believed always existed in some form, was constantly expanding, and growing like a living organism. In fact, he realized that the cosmos was the ultimate life form!

The Bible's world view as described by Genesis was useful in its time period. The old scriptures gave man the right to "subjugate" the earth, an instruction that offered a more plentiful life which man dutifully carried out with plows, sling shots and spears. Unfortunately, man had continued with this obsolete decree into the twenty-first century. Nuclear energy, cloning and global warming had placed habitation on this planet in serious jeopardy. Everywhere he looked he saw the dark cancer of man's foolish notion that mankind had the right to control the ecosystems that surround him. The air, water and soil were contaminated almost past the point of reversal. Sadly, he grew to understand that the old, God inspired cosmology of the Bible had brought mankind to the brink of a catastrophic disaster. Change was absolutely imperative or the future of not just his children, but of all life in the upcoming generations, would be in danger.

He realized that the cosmos, like mankind, was not just a living force in a constant state of evolution, but also an energy that was always seeking a more perfected expression of itself. He now understood that human beings had to invent a new cosmology that called attention to the divine energy in the physical particles that composed the universe. Only by treating matter as a complex living force, **sacred in its own right**, could mankind bring about true restoration of the earth. The old theology that emphasized the sinfulness of flesh was both obsolete and in some ways misguided.

Over the centuries, science led to the elimination of a large number of the intuitive powers with which our earlier ancestors responded to the awesome mysteries of the gigantic living body that contained us. We no longer were in awe of the heavens with its star filled night or occasional eclipse. We no longer practiced the Indian ways of "waste nothing" or "put it back." However, now our very survival depended upon green living, walking as lightly as possible on the earth while treating it as the divine organism that it is. A return to a balanced relationship with nature, understanding that its well-being directly influenced our own, was at the core of his new way of life. The truth could be reduced to one very simple maxim: treat the multitude of sacred energies that continually intersected with our lives as we would treat our most beloved. For him, the act of worship was not primarily carried out in a church sanctuary, but in the daily act of living harmoniously with the planet through the process of being nearly invisible, much like an unseen flower in a meadow or unnoticed pebble in a stream.

Occasionally he would think about his wife and two sons and feel a twinge of guilt, remorseful that their family was not fully intact. His love for them was unconditional and infinite. Every so often he would remember Grendel's tragic fate and feel the profound sadness that a parent might feel for a dead child. Mostly, however, he was able to let go of his suffering. Magically, the wound on his forehead almost totally disappeared. Only a very faint crease remained where there once was a large, reddish scar. He felt whole. As a final tribute, he attempted to live his life as if he were Grendel, mimicking in his small community as best as he could the states of consciousness that the giant primate might have experienced.

He recognized that Grendel innately manifested the majority of the traits that he endorsed in his book, *The Lustres: The Universal Wisdom of the Mystical Traditions*. Grendel lived spontaneously, moving freely through life, as a child might before being socialized. Because he lived with a natural simplicity seldom experienced by modern man, he was largely inner directed and always honoring his right brain, intuitive powers. Furthermore, he was guileless, serene and mostly non judgmental. He flowed in unison with nature, rarely interrupting its basic principles. Finally, he possessed no ego, only an instinct to survive, which allowed him at times to feel empathy for all other living beings in their quest for survival. Ultimately, Grendel was the quintessential uncarved block.

At first it was difficult to discard the traditional markers, the distinguishing attributes of the self of which his ego was so proud. Giving up the cherished roles of father, husband and son required the most time and effort. No matter how hard he tried, he never was able to fully let go of all the family ties that pulled at his emotions. Sometimes he would wake up from a nap thinking that he had to phone his mother, only to remember that she was dead. At times he wanted to send gifts to his sons on their birthdays or talk to Kathy about reconciliation. No interaction with his loved ones was better than distant words, no matter how genuine, from a person who was now a stranger.

In time he realized that since everyone inherently was divine, every interaction, even with a remote stranger, was potentially imbued with the same love as if he or she were a family member. Every older woman could be his mother and every young man a potential son. His daily discipline taught him to live as if the entire world were part of a huge, extended family, a merging of billions of pure sparks of energy, that when combined, encompassed a radiant, earthly divinity. That truly was paradise. Within reason, he would gladly help anyone who needed a hand. What he could not provide through material assistance, he could through personal involvement. Helping a neighbor build an irrigation system or working a local farm while the owners attended to an emergency accomplished his goal of giving to humanity. Every living being, the entire physical world for that matter, was just an extension of the ultimate light!

Among the dozens of other positive qualities that he associated with his former self were words like intelligent, capable, honest, caring and compassionate. But what did those lofty traits really mean? They were a pleasant reality picture that he once carried around with him. Unfortunately, the words were just as useless as a proud parent carrying in his wallet an old photograph of his adult children. Why be preoccupied with a world that no longer existed? When he realized that he had become attached to those pretty images, as others were attached to a new car or house, he worked with even greater diligence to let go…to fully embrace the universal flow. Intelligent or honest compared to what? He knew that if he were to continue to remain fully alive, he had to move beyond these neatly crafted illusions that he had imposed on the world with the help of his self indulgent culture.

The only state of being worth striving for was the powerful, surging light of **now**, beyond the illusionary dualism produced by the allure of words and comparisons.

Often his mind would revert back to a haunting image that he saw years ago while watching a movie that was popular at the time. The opening scene of the film seemed quite ordinary, a slow zoom through thick, iron bars. Was it a prison or perhaps a zoo? Then, a figure began to emerge from the darkness and gradually clarified. In the back of the cage, almost indistinguishable from the gray shadows, a large, old gorilla sat serenely eating a banana with one hand and scratching his chest with the other. Amazingly, the cage door was wide open and there was no zookeeper in sight. For all practical purposes, the gorilla was free and if he wanted, he could experience the joys of life beyond his dark confinement. Yet he sat absolutely motionless, almost as if he were suffering from a form of paralysis, only occasionally chomping on his banana. He wondered, *how could the zookeeper's bananas ever be enough to provide real happiness?*

For a man attempting to liberate himself, this was a depressing metaphor, one that quite accurately represented the predicament of most people in contemporary America. Many of his former students readily confessed that they felt their lives, like the fate of the old gorilla's, were already pre-constructed for them by their parents and their culture, as if they all shared the same blue print for an upscale model house in the same prestigious suburban development. They were taught that all they had to do was physically show up and everything, including a successful career, would take care of itself. How frightening! More than anything he did not want his day-by-day world to erode into such a pitiful state of stagnancy, one where he had become so over socialized that he was totally indifferent to the actual quality of his immediate life. *How could a cage, even if constructed out of the purest gold, ever be preferable to genuine existence? Who could be satisfied with simply showing up and being placed in an artificial, plastic mold?*

So over the years, with the application of the ten lustres from his book, he gradually disappeared. He had a healthy physical embodiment yet he played no part in his culture's world of routine and posturing. Mostly he went unnoticed. Because nobody understood his life, few people, if any, paid attention to his vibrant inner being. Often, they felt good in his presence but they never stopped to consider why.

Because he lived very simply, following the flow of his natural energy, some might refer to it as his soul, he took no interest in the money driven culture that others were obsessed with. He did not concern himself with the stock market or the latest, end-of-the-year clearance sale. Each morning he arose from a night of restful sleep and then promptly disappeared into a deep meditation, with Tao, his two year old black lab, settled comfortably at his feet. His bond with Tao was that of two soul mates: sacred, egoless energies openly and fully embracing one another. As the sun rose over the glorious hills, with perfect serenity he silently recited his mantra, something like "celebrate this very moment for this very day, **this** is paradise" or the Jesus prayer. After a few moments the words automatically repeated themselves in his brain and joyfully his empty head floated effortlessly upward into another realm. Some days he even savored the miraculous strawberry that revealed itself deep within the center of his forehead, just between his eyes.

For him, life was not endless conflict, not a series of battles to be dominated by his will power. Authentic existence meant drifting towards transcendence, experiencing the moment as a glorious inner light, a radiant, orange ball of sunshine, rhythmically dancing within his being. He lived in that sacred space much of the day. Others worked at jobs that they complained about to earn money to buy things that they didn't really need to impress people that they didn't really know or care about. He lived quietly while engaging in numerous solitary pursuits: contemplating, reading, meditating, and writing, but more often than not, walking and playing with Tao. All the while, his soul was singing happy songs that nobody heard, melodies so powerful that, at times, he felt connected to the Absolute. At last his actions were mostly egoless. If a neighbor were to observe him, detached from the outcome of his worldly involvement and with a distant look in his eyes as well as a joyful smile on his lips, it might appear that he was not there. There was a physical form dressed in his clothing. But most people, especially those attached to ordinary reality, could never fully grasp the idea of liberation. Even if they wanted to interact with him, nobody quite knew how to talk with someone so fully present, so totally free.

People did not usually invite him to the community social gatherings because they could tell that being around dozens of people, all artificial in some manner, made him uncomfortable. He had no desire to be standing by a pool with a drink in his hand, listening to an inane joke or talking about an utterly trivial matter. Besides, he was different. He was happy just being himself. Either he didn't know how to "act" around people or he knew how to perform but chose not to. The result was the same: even when he was there, he seemed in the eyes of others to be in some far away place.

On rare occasions, a few people would take the time to meaningfully converse with him, not about the latest golf tournament or a new blockbuster movie, but about real issues that related to their lives. Often they walked away uplifted, amazed by his wisdom. However, the most successful people were put off by him, wondering, "How could he waste his life like that? What was the value of not competing? What kind of person wanted to live on nothing?" So they avoided him. But in the great scheme of things it didn't matter. Their luxury and his simplicity would, in time, come to the same end, only the impact on the planet might be different.

He had learned from many religions, but especially from the Native Americans, another way of understanding life. "Leave no trace" and "Put everything back" resonated with his deeper self. Most especially, he wanted to "find a pathway with a heart." While most of his neighborhood was striving to carve out their place in the world, he lived in harmony with a higher order of things, feeling a truly powerful bond to both Tao and the continuous bounty of Mother Earth. Now, he spent his days innocently celebrating the often taken for granted gift of life. He had realized long ago that it was easier for almost everyone if he were invisible, so he continued to cultivate a natural life that few people, especially Kathy, really understood.

In the summertime, he would tend to his bountiful garden, sweat poring from him as he meticulously cared for each plant. He enjoyed the fruits of his labor-eating tomato and onion sandwiches on home-baked bread or stir-fried vegetables over wild rice. He was most happy when his body, stripped to the waist, glistened in the morning sunshine. Physical labor always felt cleansing, as if his skin were a golden sheen of untarnished energy. Sometimes as he worked the soil, he remembered with sadness Grendel's peaceful nature and pure animal spirit.

Often he took long, rambling walks with Tao into the nearby hillside, sometimes following a path, but just as often bushwhacking, heading in whatever direction his fancy took him. All of his energy was concentrated in the moment. Sometimes, he simply searched for the source of a dried up stream or with Tao, basked in the sunlight, perched on a protruding rock high above the village below. At that moment he sometimes pretended that he was Grendel secretly peeking at creatures who voluntarily incarcerated themselves in diminished realities.

Most importantly, each day he sampled the magical taste of utter freedom, the bliss of flowing with the timeless, living universe. For many people, his life would be unimaginably boring and, perhaps, hopelessly lonely. Yet for him the joyful energy of each moment embraced his spirit as a most beautiful lover united with his beloved. It was at that magical moment that he was most invisible-nameless, ageless, genderless and even in a sense formless. He was identical to Tao and Grendel: an egoless, surging natural energy. Of course there was a body and clothing, but in the truest sense of the word, it was not his. Everything physical was simply borrowed from nature and over time returned. He knew that most people unknowingly adopted an identity and then spent everyday pretending to be a business man or professional. Years ago he was consumed by the same false game playing, but the suffering became unendurable.  On occasion he thought about going back to father, husband, provider, but those roles would just cause more misunderstanding for those he loved.  He could not effectively juggle a series of masks simply to make others happy, no matter how much he cared for them.  *To mindlessly repeat the same day over and over again, now what healthy, vibrant person would chose to do that?*

With Tao, time passed in serene, inward evolution, and not a frantic effort to find the fountain of youth or even what some might call his "true self."  As his life unfolded, the inner gates of his being spontaneously opened to the bliss of unimaginable beauty and awareness; gradually he learned to be completely still and just observe.

He enjoyed the integrity of being a mostly uncarved block.  When he no longer analyzed life, he began to not only see, but actually feel the real perfection of everything around him.  For what seemed like hours he could watch tiny insects crossing the symmetrical veins on the underside of a pale green leaf or hawks, with dark wings extended, gracefully floating in the pure blueness of a mid-morning sky.

Everywhere and in everything he felt a sense of wonder.  He lived on the edge of ecstasy, connected with life's radiant holiness.  And in certain, unexpected moments the beauty of it all was almost unbearable.  He was no longer a role, job description or social construction of some well intentioned culture, but simply an undefined energy, almost completely invisible, even to himself.

In those moments, he simply did not exist, at least as our society thought of existence.  And it was then that he did not just hear the divine melody; he became the Song, the Dance, the Light, inseparable with the Composer… his ears listened to the gentle surging of water falling melodiously onto a bed of stones…his skin felt the warmth of the morning sun, dancing joyfully on his arms and face…his eyes watched an ocean of flowers tilt their colorful blossoms towards the changing light, playing hide and seek with the shade… he experienced everything without judgment…he became a mindful witness, free of intellectualization and the oppressive weight of the ego…he and Tao seemed identical.

The universe existed and he, for the moment, was part of it…he breathed in and out, enjoying the bountiful splendors of deliverance.  At last, he was fully alive, totally free and completely invisible, far beyond words and the painful struggle of human comprehension.  He came to perceive all of creation as nothing but a fantastic symphony of light, thousands of diverse manifestations, each an individual spark, and all composing a sacred, omnipotent radiance, a warm glowing joyfulness that was as simple as it was eternal.